The Evolution of Global Crustal Uplift and Depression

Changes of Sea and Land

The Evolution of Global Crustal Uplift and Depression
Changes of Sea and Land

Yuzhu Kang
Sinopec Petroleum Exploration and Production Research Institute, China

Zhihong Kang
China University of Geosciences (Beijing), China

Zhijiang Kang
Sinopec Petroleum Exploration and Production Research Institute, China

Jiwei Wang
Sinopec Petroleum Exploration and Production Research Institute, China

World Scientific

NEW JERSEY · LONDON · SINGAPORE · BEIJING · SHANGHAI · HONG KONG · TAIPEI · CHENNAI · TOKYO

Published by

World Scientific Publishing Co. Pte. Ltd.

5 Toh Tuck Link, Singapore 596224

USA office: 27 Warren Street, Suite 401-402, Hackensack, NJ 07601

UK office: 57 Shelton Street, Covent Garden, London WC2H 9HE

Library of Congress Cataloging-in-Publication Data
Names: Kang, Yuzhu, author. | Kang, Zhihong, author. | Kang, Zhijiang, author. |
    Wang, Jiwei (Petroleum engineer), author.
Title: The evolution of global crustal uplift and depression : changes of sea and land /
    Yuzhu Kang (Sinopec Petroleum Exploration and Production Research Institute, China),
    Zhihong Kang (China University of Geosciences (Beijing), China),
    Zhijiang Kang (Sinopec Petroleum Exploration and Production Research Institute, China),
    Jiwei Wang (Sinopec Petroleum Exploration and Production Research Institute, China).
Description: New Jersey : World Scientific, [2024] | Includes bibliographical references and index.
Identifiers: LCCN 2023049053 | ISBN 9789811286063 (hardcover) |
    ISBN 9789811286070 (ebook for institutions) | ISBN 9789811286087 (ebook for individuals)
Subjects: LCSH: Earth (Planet)--Crust. | Geodynamics. | Geology. | Plate tectonics.
Classification: LCC QE511 .K26 2024 | DDC 551.8--dc23/eng/20240129
LC record available at https://lccn.loc.gov/2023049053

British Library Cataloguing-in-Publication Data
A catalogue record for this book is available from the British Library.

全球地壳隆坳演化与海陆变迁论
Originally published in Chinese by China Petrochemical Press Co., Ltd.
Copyright © China Petrochemical Press Co., Ltd. 2020

For any available supplementary material, please visit
https://www.worldscientific.com/worldscibooks/10.1142/13672#t=suppl

Desk Editors: Nimal Koliyat/Julio Hong

Typeset by Stallion Press
Email: enquiries@stallionpress.com

Preface

In the past century, geologists have carried out a lot of research on geodynamics and the crustal evolution of the earth, having formed many schools of thought and theories, such as the continental drift theory and the plate tectonics theory. The author has come to believe, after decades of research, that there are in total seven continents on the crust around the world, namely, Eurasia and the continents of Africa, North America, South America, India, Antarctica, and Australia. The crust has covered the surface of the earth for billions of years since its formation and has been an integral part of it. It has never been separated from the earth in the past, nor will it be separated in the future.

Since the existence of human beings on the planet, there has never been crustal cracking but instead faults, volcanic eruptions, and magmatic intrusions in different areas, toward different directions, and of different properties and scales that occurred locally in the continental crust and marine crust, which will continue to occur in the future. These geological events can cause earthquakes of varying intensities and local crustal deformation, such as mountain- building by crustal uplift or depression by subsidence, due to the compression or extension activities by local geostress.

This book was completed with support from many officials from the Ministry of Land and Resources, China Geological Survey, the Southern Marine Science and Engineering Guangdong Laboratory (Guangzhou), and the Research Institute of Petroleum Exploration and Development of the CPCC. In particular, the guidance of Shuwen Xing and Long Changxing, Directors of the Institute of Geomechanics, Chinese Academy of Geological Sciences, has been much appreciated. Many works published by geologists, oil and gas geologists, and other scientific researchers have been referenced in this book. The author would like to express his heartfelt thanks to the individuals who wrote those books. In the process of writing this book, the author also received financial support from the Key Special Project for Introduced Talents Team of the Southern Marine Science and Engineering Guangdong Laboratory (Guangzhou) (GML2019ZD0102), and hereby expresses his gratitude.

If anything improper is found in this book, please do not hesitate to point it out.

List of Authors

Yuzhu Kang, Academician, Sinopec Petroleum Exploration and Production Research Institute, China Petrochemical Corporation, Beijing, China

Zhihong Kang, China University of Geosciences (Beijing), Beijing, China

Zhijiang Kang, Sinopec Petroleum Exploration and Production Research Institute, Beijing, China

Jiwei Wang, Sinopec Petroleum Exploration and Production Research Institute, China Petrochemical Corporation, Beijing, China

Shihe Liu, Project Editor, General Editorial Department, China Economic Publishing House, China Petrochemical Corporation, Beijing, China

Hongyu Song, General Editorial Department, China Petrochemical Press, China Petrochemical Corporation, Beijing, China

Contents

Chapter 1

The Origin of the Earth's Motion

Abstract

This chapter discusses the origin of the earth's motion. Changes in the earth's rotation speed are an important driving force for its movement. This chapter analyzes the influence of the earth's rotation and impacts of celestial bodies on this rotation in terms of internal causes from the earth itself, geostress caused by the difference in thickness and density of the crust, and the earth's layer structure.

Keywords: Earth, motion, driving force, constitution.

1. Effects of the Earth's Rotation

Mr. Siguang Li [1], a celebrated Chinese geologist, pointed out long ago that changes in the rotation speed of the earth are an important driving force for its movement. The angular momentum of a rotating object is constant, which in general is expressed as follows:

$$wI = C,$$

where W stands for the angular velocity of a rotating object, I is the moment of inertia of the rotating object around its axis of rotation, and C is a constant.

When I changes, w will certainly change in inverse proportion: When I decreases, w will increase. When the mass of the earth moves toward its center, I will, beyond doubt, decrease. Such a change may originate from

1

several different actions: (i) shrinkage of the earth as a whole (the Shrinking Earth Theory); (ii) massive subsidence of the crust (the Vertical Motion Theory); and (iii) gravity differentiation that may occur inside the earth and convection of lava with different densities. No matter which of these three hypotheses is close to the reality, as long as one of these actions, or at a certain stage, makes the mass of the earth converge by a certain degree toward its center, the angular velocity of the earth will increase to the extent that its shape as a whole will change. When the force resisting such changes on the surface of the earth or at the upper layer of the crust is less than the force inside, especially when the geoisothermal surface rises, a horizontal force of certain intensity will easily produce a pushing effect on the upper layer of the crust, thereby causing the earth to form a new shape. It goes without saying that the force resulting from this action is the horizontal component caused by the combined action of the increasing centrifugal force and gravity when the earth accelerates its angular velocity. This horizontal component [2, 3] meets the demand for horizontal movement of certain portions of the earth's crust, especially the demand for the formation of an epsilon-type structural system.

Meanwhile, the crust, or its upper layer, is not necessarily uniformly fixed to its base. If the two adjoining parts of the earth's crust do not accelerate at the same pace along with the rotation speed of the earth, there will be compression and tension cracks [4] in the E–W extension. If a part on the east side does not accelerate with the rotation of the earth like a part on the west side does, there will be compression and torsion cracking on the horizontal plane along the N–S extension between the two parts. Likewise, if a part on the west side does not accelerate with the rotation of the earth like a part on the east side does, there will be tension cracking and torsion cracking on the horizontal plane along the N–S extension. In this case, due to the change of the angular velocity of the earth, the two fractures along the N–E and N–W extensions can produce not only an E–W extending tectonic system and an epsilon-type tectonic system but also an N–S extending tectonic system.

Based on the principle of conservation of angular momentum, when the angular velocity of the earth decreases, the moment of inertia around its axis of rotation should increase, that is, its mass distribution should spread outward, its volume should increase, or the materials with less density should move upward to the surface of the earth on a large scale. The view that the change of the earth's moment of inertia causes the change of angular velocity was put forward in China and Hungary

(Schmidt) at the same time more than 30 years ago, which cannot be regarded as coincidental. A large number of basaltic flows developed during the Permian in southwestern China and other regions of the world, and Deccan dark rocks spanning an area of about 100×10^4 km^2 were exposed in the Indian Peninsula in the early Paleogene. Basic rock flows widely distributed in many regions of the western Indian Ocean, the northern Atlantic Ocean, and the Pacific Ocean, as well as various dense igneous rock batholiths and rock masses intruding into the upper part of the crust that formed during different large-scale orogeny eras, all carry historical traces of the large-scale rise of the dense materials below the crust or in the lower part of the crust [5].

When the mass distribution of the earth changes and its angular velocity becomes smaller under the influence of tidal action, the flatness of the earth will be too large to meet the demands of its rotation speed. In this case, faults and folds along the E–W and N–S directions may occur. Then, has the angular velocity of the earth ever changed? Records of ancient solar eclipses and observations of several astronomers in modern times gave an affirmative answer to this question. Most scientists believed that the angular velocity of the earth had a general tendency to slow down, while others like Urey held the opposite view. In fact, historical records prove that the speed of the earth's rotation is sometimes fast and sometimes slow. Among its speed changes, there is a kind of "irregularity" of becoming faster and slower. Although the extent of such irregular changes is small in historical times, we do not use such historical epochs as a measure of possible changes in geological times. In other words, we have no reason to rule out the possibility that the cumulative changes in the rotation speed of the earth during the geological age sometimes exceeded the critical value at which the shape of the earth's surface can maintain a balance [6]. Therefore, we have no reason to conclude that the earth stopped its mass concentration before the movements of its crustal surface reached the critical state.

2. Effects of Celestial Bodies on the Earth

The reasons for the change in the angular velocity of the earth and the underlying conditions for the gradual directional movement in and under the crust of the earth may come from celestial bodies closely related to the earth, especially the moon and the sun. From the perspective of the requirements for the directional movement of the crust, some astrophysicists (such as Taylor, Jolie, and Lichkov) thought that the tidal effects of the moon on the earth were the general reason for tectonic movements in the earth's crust. Several astronomers and astronomical geologists from the former Soviet Union put forward arguments about the assumption that the sun's activities might affect the movement of the crust. Among them, some, such as Egenson, believed that the sun's activities might affect the angular velocity of the earth, and others, such as Snarski, believed that the sun's activities were related to the changes in the intensity of the magnetic field of the earth. By making use of the hypothesis that when the magnetic field weakens, substances free from magnetism generate heat and the fact that the 11-year changing cycle of the magnetic field of the earth corresponds to the 11-year cycle of solar activities, Snarski proposed that the rise and fall of the isogeothermal surface of the earth's crust was not the cause but the consequence of the change of the magnetic field intensity [7]. This novel hypothesis proposed that changes in the magnetic field intensity of the earth created conditions in the crust that made tectonic movement possible.

3. Reasons from the Inside the Earth Itself

Radioactive substances are widely distributed in the earth's crust and they constantly generate heat. When the temperature on the earth's surface is roughly constant and the rock heat transfer rate and geothermal gradient remain constant, the temperature in the lower part of the earth's crust is likely to gradually increase. Based on this possibility, Jolie concluded that the rocks in the lower part of earth's crust would melt approximately every 30 million years. In his book *A Theory of Earth's Origin*, Schmidt made great progress in the study of the importance of radioactive materials in the earth's thermal history. Many geologists, including Holmes who initially used the radioactivity of minerals to identify the earth's age, did a lot of work in this area. It seems that the existence of radioactive elements in the crust is closely related to its thermal state. However, how radioactive elements are distributed in all parts of the earth's crust and even below the crust is still an open question. It is unreliable to conclude on the distribution law of radioactive materials in and below the crust based on the radioactivity of several types of rock specimens alone. As Kraskovski pointed out, different kinds of data about the earth's thermal state measured around the earth in the past have not been reliable [8].

Thus, the degree of isothermal surface changes proposed by Jolie and other geologists based on the assumption that radioactive elements are distributed according to a certain law in the earth's crust needs to be closely evaluated and studied. This effect mainly creates conditions for the directional movement of the earth's crust. Therefore, temperature difference inside the earth can cause local movement of the earth's crust.

4. Geostress Resulting from Differences in Thickness and Density of the Crust

Due to differences in crustal thickness and density, compressive and tensile stresses are formed during the rotation of the earth, which cause various structural deformations. In the process of rotating, the regions with large crustal thickness and high rock density produce compressive stress on the regions with a thin crust and low rock density, thereby causing corresponding deformation.

5. The Spherical Layers of the Earth

The matters inside the earth are differentiated by gravity, with the heavy matter at the bottom and the light matter at the top, forming the spherical layers of the earth. The thickness and the density of each layer are shown in Figure 1 and Table 1. There are two spherical layers, the densities of which vary the most. One lies in between the mantle and the core, with the density difference being an order of magnitude. The density difference of the other spherical layer is much bigger, located as it is in between the atmosphere and the crust, with the difference being three orders of magnitude. The differentiation between the mantle and the core and the generation of the atmosphere and ocean, as discussed in the following, lead to the most significant variations in the formation process of the spherical layers of the earth [9].

There is stratification inside each spherical layer of the earth. The core of the earth is mainly composed of iron and nickel. The density of the core is as high as $9.7–16$ g/cm^3, making the overall density of the earth greater than 5.5 g/cm^3. The earth is the planet with the largest density in the solar system. The core of the earth is divided into inner and outer parts. It is presumed that the inner core is in a solid state while the outer core is in a

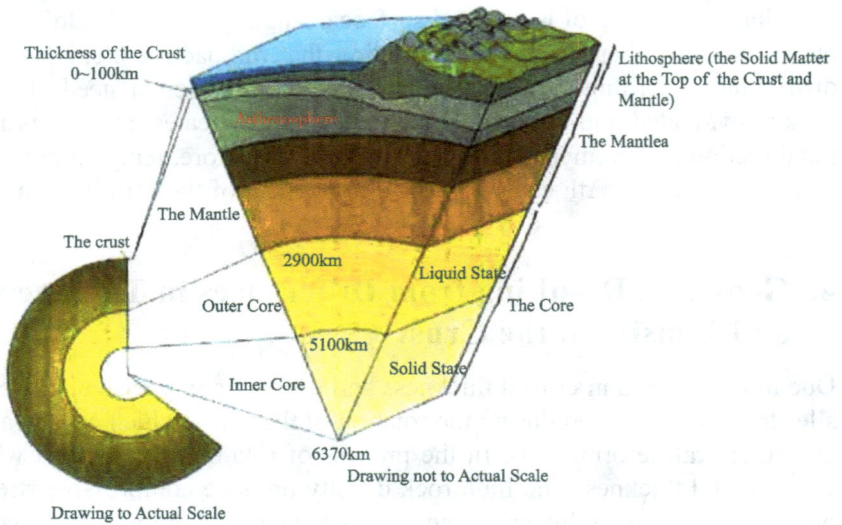

Figure 1.   The spherical layers of the earth (adapted from Wikipedia).

Table 1.   Thickness and density of each spherical layer of the earth.

Items	Inner core	Outer core	The mantle	The crust	Hydrosphere	Atmosphere
Thickness/km	1,200	2,300	2,860	35	4	700
Density/(g/cm³)	12.6~16	9.7~12.2	3.3~5.7	2.7~2.9	4	≤10⁻³

liquid state. The temperature of the outer core is over 4,000°C and that of the inner core is over 5,000°C, as high as the temperature of the solar surface. The mantle, composed of silicates of iron and magnesium, is divided into upper and lower layers. The thickness of the upper layer is 400 km and that of the lower layer is 2,200 km. There is a transitional layer between the upper and lower mantles with a thickness of 300 km. The top of the upper mantle and the crust combine to form plates that participate in plate movement, which is the object of the geological research on tectonic movement; these plates form the lithosphere, with a thickness of about 100 km. Different from the overlying lithosphere, the upper mantle, named asthenosphere with a thickness of 300 km, is in a plastic state and can be viscously deformed. In the transitional layer of the

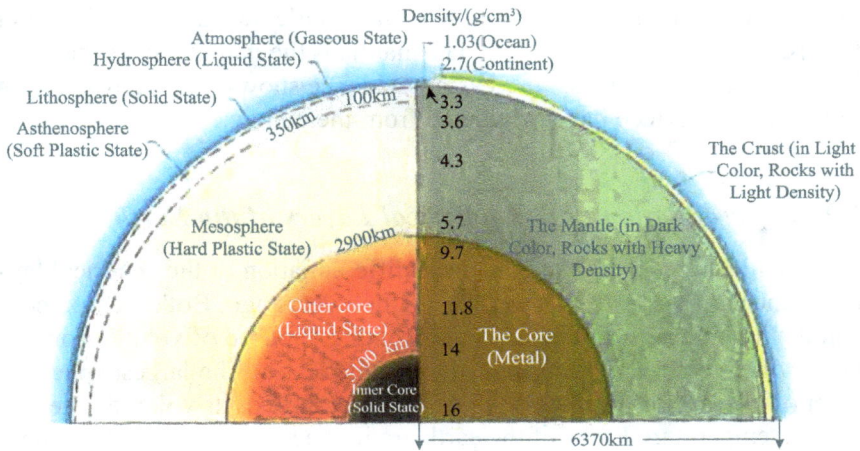

Figure 2.   Thickness and density of the spherical layers of the earth (adapted from https://www.studyblue.com).

middle section of the mantle, which is the location of the deepest seismic movement, seismic wave velocity suddenly increases. In the lower mantle, the pressure increases and the seismic wave velocity accelerates. The bottom layer, called layer D, is directly affected by the core material and plays an important role in mantle cycling [10]. The crust of the earth is more familiar to humans. Basalt oceanic crust and granitic continental crust are different in thickness and structure (Figure 2). The oceanic crust is generated in the mid-ocean ridge. The basalt oceanic crust, formed by upwelling magma, expands to both sides while cooling, and finally disappears in the fault subduction zone and returns to the mantle.

5.1. *The Mass of the Spherical Layers of the Earth*

Roughly speaking, the volume of the earth is more than 1 trillion cubic kilometers and the mass is close to a hundred million times 60 trillion metric tons. The "surface system" that human beings come directly in contact with is tiny. On this scale, the mass of the atmosphere and hydrosphere is so tiny that it is negligible. In the "inner spherical layers" of the earth, in terms of volume, the core accounts for 16.2%, the mantle accounts for 83%, and the crust accounts for less than 1% only; in terms of mass, the core accounts for 32.5%, the mantle accounts for 67%, and the crust accounts for 0.5% only. The mantle is the main body of the earth by both volume and mass. Increasing evidence shows that many changes in the earth's surface have originated from the mantle.

5.2. *Differentiation of the Spherical Layers of the Earth*

Geochemical evidence shows that the differentiation of the spherical layers of the earth occurred around 4.45 billion years ago. Both the iron core and the hydrosphere and atmosphere, being composed of volatile components, diverged at that time from the magmatic sea. The largest spherical layer of the earth is the mantle, accounting for 84% of its volume. The two largest temperature limits of the earth are in the mantle: in the lithosphere at the top of the mantle and the so-called D layer at the bottom of the mantle. Therefore, the differentiation of the magmatic sea was attributed primarily to the formation of the mantle. How the core of the earth, which is close to one-third of the mass of the earth, separated from the mantle is the primary subject in the analysis of the differentiation of the spherical layers of the earth.

5.3. *Formation of the Core, Mantle, and Crust of the Earth*

The formation of the core and mantle was essentially caused by the differentiation of siderophile elements, such as iron, nickel, cobalt, and manganese, and lithophile elements, such as silicon, aluminum, magnesium, and calcium. The chemical composition of the earth before the differentiation would have been similar to that of the current carbonaceous chondrites. Compared with the chemical composition of the mantle and crust of the current earth, the ratio of the siderophile elements and the lithophile elements was much higher and the Fe/Al value was as high as 20. The Fe/Al ratio in the upper mantle of the current earth is 2.7, and only 0.6 in the crust, because the siderophile elements are concentrated in the core [11].

The earth was formed by the aggregation of a large number of planetesimals (accretion bodies) over a period of 30 million years. Each planetesimal contained two kinds of components represented by iron and aluminum/silicon. Even under the high temperature and pressure in the earth's interior, like oil and water, the two components could not come together. How exactly the iron brought by planetesimals was incorporated into the core of the earth has aroused much debate in academic circles over the years. It has now been inferred that the original upper part of the earth was a magmatic sea, whereas the interior of the earth was solid. Planetesimals collided and merged into the earth. The lighter silicon–aluminum remained in the upper layer, while the heavier metal "droplets" descended through the molten silicate layer, stopped at the bottom of the magma sea at a depth of about 400 km underground, and gathered into a metal "pool." When this metal layer became unstable, it descended to the center of the earth in the form of large droplets, forming the core of the earth (Figure 3). As of today, humans only have an indirect understanding of both the core and mantle of the earth, and it is unclear whether the abovementioned differentiation was a complicated and repeated process. The current understanding is that, after the Theia Impact, the metal layer in the middle mantle suddenly sank, which triggered a cataclysm of core–mantle separation and constituted the first and most important reorganization event [12].

The three major spherical layers inside the earth, i.e., the crust, mantle, and core, are vastly different in densities because of their different

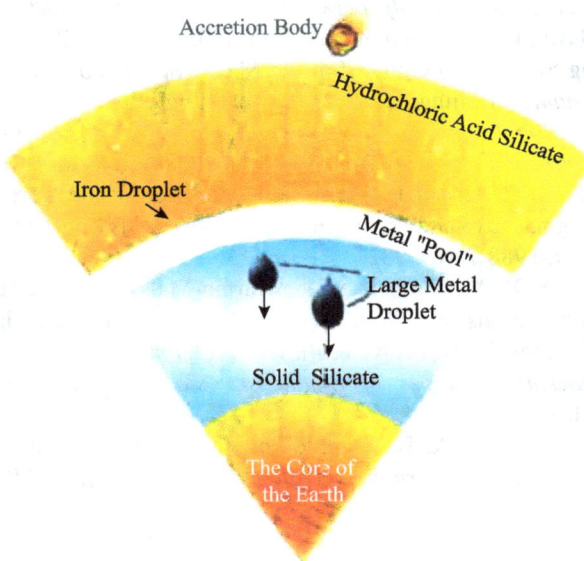

Figure 3. Aggregation of the siderophile elements in the formation of the core of the earth.

compositions. During the rotation of the earth, the velocities of these three spherical layers are different, and a new geostress is generated accordingly.

References

[1]   Siguang Li. *Methods of Geomechanics*. Beijing: Science Press, 1976.

[2]   Songsheng Li. New understandings on arc tops of Huaiyan epsilon-shaped tectonics. *Proceedings of 562# Comprehensive Brigade, Chinese Academy of Geological Sciences*, 1991 (9): 37–45.

[3]   Jitao Liang, Debao Zhang. Survey on China ancient continent. *Journal of Nanjing Institute of Geology and Mineral Resources, Chinese Academy of Geological Sciences*, 1991 (2): 3–12.

[4]   Baojun Liu. *Crustal Evolution and Mineralization of Ancient Continents in Southern China*. Beijing: Science Press, 1993.

[5]   Yuzhu Kang. Petroleum geology features and petroleum prospects in Northwest China. *Petroleum Experimental Geology*, 1984 (3): 72–83.

[6]   Yuzhu Kang. Discovery of high-yield oil and gas flow in Shashen No. 2 well and future prospecting direction. *Oil and Gas Geology*, 1985 6(S1): 45–46.

[7]   Yuzhu Kang. *Relationship between Tectonic System and Oil and Gas in Tarim Basin*. Beijing: Geological Publishing House, 1989.

[8]   Dianqing Sun, Peishi Gong, Zongjin Ma, *et al. Methods and Practices of Geomechanics*. Beijing: Geological Publishing House, 1997.

[9]   Gill, S. *Global Structural Change and Multilateralism. Globalization, Democratization and Multilateralism*. UK: Palgrave Macmillan, 1997.

[10]   Aadland, R. K., Schamel, S. Mesozoic evolution of the Northeast African shelf margin, Libya and Egypt. *Bulletin American Association of Petroleum Geologists*, 1988 72(8): 982.

[11]   Aeharyya, S. K. Mobile belts of the Burma-Malaya and the Himalaya and their implications on Goudwana and Chthaysia Laurasia Continent Configurations. In: N. Prinya (ed.), *Third Regional Conference on Geology and Mineral Resources of Southeast Asia* (pp. 121–127). Bangkok, Thailand, 1978.

[12]   Achnin, H., Nairn, A. E. M. Hydrocarbon potential of morocco. *AAPG Bull, Association of Petroleum Geologists*, 1988 72(8): 32–39.

Chapter 2

An Overview of the Sedimentary System of the Crust

Abstract

This chapter discusses crustal sedimentary systems, including the Cambrian–Middle Ordovician, the Upper Ordovician–Siluric–Devonian System, the Lower Carboniferous Series, the Upper Carboniferous–Lower Permian Series, the Middle-Upper Permian Series, the Triassic–Lower Jurassic Series, the Middle Jurassic–Lower Cretaceous Series, the Cretaceous System, and the Tertiary System.

Keywords: Sedimentation, Ordovician, carboniferous, Jurassic.

Over the years, experts and scholars have conducted a lot of research on sedimentary characteristics all over the world. On the basis of these existing studies, the author provides an overview of sedimentary systems all over the world from the Palaeozoic.

Three major sedimentary systems have been discovered globally: the marine sedimentary system, the marine–terrestrial interaction sedimentary system, and the terrestrial sedimentary system. However, these sedimentary systems vary a lot between the major landmasses and within landmasses, in different eras, and in the different basins. Thus, the features of sedimentary systems can only be summarized macroscopically.

1. Cambrian–Middle Ordovician

From the Late Proterozoic to the Early Cambrian, the Rodinia supercontinent changed with the Baltic, Siberian, and Laurentian continents, moving northward. The Baltic and Gondwanan continents developed a combined continental shelf and passive continental margin. As such, extensive marine carbonate platforms developed in the Laurentian, Antarctic, Chinese, and Siberian continents, whereas clastic rocks were relatively limited. During the Middle-Late Cambrian, the Siberian and Laurentian continents continued to move northward, and the local extension resulted in the development of inland shelf basins on the passive continental margin of the Laurentian continent. In the Early Ordovician Period, continents all over the world continued to change, resulting in convergence along the southwestern margin of the Laurentian continent and the northwestern margin of the Baltic continent. The Cambrian–Middle Ordovician included a first-order sequence, which was divided into five second-order sequences [1]. In the Early Cambrian, the glacial climate prevailing in the Late Proterozoic gradually transited to a warm climate. During the Middle Cambrian to Middle Ordovician, a humid climate was widespread.

In the lower Cambrian, gentle slope-type carbonate platforms developed, which were composed of high-energy oolitic shoal, land-ward low-energy lagoons, tidal flat deposition, and sea-ward organic-rich mudstone and marl. From the Middle Cambrian to the Middle Ordovician algae, foraminifera and oolitic carbonate platforms developed. During the Early Middle Cambrian a first-order rise in sea level occurred, leading to widespread intrusion of seawater into cratons. Marine carbonate platforms widely developed in the continents of Laurentia, Siberia, Eastern Antarctica, Australia, and China. When it came to the Middle-Late Cambrian, a rise in sea level caused further flooding. As a result, extensive carbonate platforms developed in Australia, China, Laurasia, and Siberia, and smaller combined platforms developed in Eastern Antarctica, Africa, and South America. During the Late Cambrian to the Middle Ordovician, a drop in sea level caused the discontinuation of marine carbonate deposits in Australia, Eastern Antarctica, Africa, and South America and reduced the scale of carbonate platforms in China, Siberia, and Laurasia, resulting in the development of large-scale clastic rock formations. Development of carbonate platforms in the Baltic area started in the Late Cambrian–Middle Ordovician. During the Early-Middle Ordovician, large-scale evaporite sediments developed in the platform areas of Siberia, Baltica, China, and North America as a result of global regression and humid climate conditions.

2. Upper Ordovician–Siluric–Devonian System

In the Late Ordovician, the separation of the Baltic, Siberian, and Laurentian continents stopped, whereas new movement and convergence activities occurred along the southern and eastern margins of the Laurentian continent and the northwestern margin of the Baltic, resulting in landmass recombination and the formation of two main continents: Gondwana (from the South Pole to the Equator) and North Pangea (including the Euramerican and Siberian massifs) [2]. The main tectonic activities during that period also included the early activities of the Hercynian and Caledonian orogeny.

In the Early Devonian, the Siberian continent separated from the European and American continents along the northern edge of the North American continent. At the same time, Gondwana began to move rapidly to the south and east margins of the European and American continents, resulting in subduction margins on both sides of the Rick Ocean. At the end of the Devonian, the Rick Ocean closed, the continents of Africa, South American (including northwest Gondwana), and Euramerica collided with each other to build mountains, and the passive continental margins were distributed along the western margin of North America and Verkhoyansk. The Upper Ordovician–Upper Devonian Series indicates that there were four second-order super-sequences in the early stage of the global first-order sea level rise in the Palaeozoic. From the Late Ordovician to the Early Silurian, the cold climate prevailed, while in the Silurian and Devonian, warm climate prevailed all over the world.

In the Late Ordovician, sub-tidal mud and spherulitic dolomite developed. During the Devonian, granular limestone, reefs and platforms, and tidal flat dolomites developed. During the Late Silurian–Devonian Period, warm temperature prevailed all over the world and karstification had a strong effect on the improvement of reservoir porosity and permeability. In the Late Ordovician, vast craton landmasses (Baltica, Siberia, North American, and Gondwana) developed isolated on a global scale, forming a wide continental shelf system, inside which inner craton depressions and basins developed, particularly in North America and Siberia. Silurian carbonate rocks mainly developed in North America and Siberia, and small-scale carbonate platforms also developed in the northern Baltic and the east of South Gondwana. In the Early Devonian, the global decline of carbonate yields and long-term sea level decline led to limited development of carbonate rocks, which were mainly distributed in the narrow strips of northern Canada, Siberia, and the east and south of the United States (the Permian Basin). During the Mid-Late Devonian, there was large-scale transgression all over the world, which resulted in prosperous reef platforms. Expansive carbonates characterized by isolated plateaus, porous shelf margins and reef-bearing craton depression were mainly deposited in western Canada, the Timan-Pechora Basin near the East European continent, the Volga–Ural region, the periphery of the Caspian Sea, and Southeast Siberia. Small-scale carbonate deposits developed in northern-central Europe, the Canning Basin in Australia, southern Kazakhstania, northeastern United States (the Appalachian Basin), and southeastern Canada. From the Upper Ordovician to the Upper Devonian, the development of clastic rocks was relatively limited across the world [3]. Affected by the Caledonian Movement, the North China platform uplifted as a whole, resulting in the development of denudation.

3. Lower Carboniferous Series

In the Early Carboniferous, the continents of Africa, South America (including northwest Gondwana), and Euramerica converged and the Rick Ocean closed. The intense compression not only affected the deposition of the Rick Trough but also controlled the development of foreland basins on the edge of cratons around the Rick Ocean. At the same time, the remaining continent of China (including South China, Indochina, and North China) rifted and separated from East Gondwana and passive continental margins developed on the northeastern edge of Gondwana. The subduction margins along the arcs of the West Siberia and Ural volcanoes extended through the Caspian Sea and the Black Sea and entered the foreland tectonic environment with strong development of platforms in east Europe and west Siberia.

The Lower Carboniferous Series was composed of 14 second-order super-sequences that can be globally compared. It was a transitional period starting from the ice chamber climate prevailing in the Devonian to the greenhouse climate of the Pennsylvanian Period. Affected by large-scale biological extinction during the Frasnian–Famennian Crisis in the

Late Devonian, there was essentially no reef in the Mississippi carbonate platforms.

Controlled by the Late Devonian extinction crisis and the global cool climate, carbonate rock deposits during the Lower Carboniferous were mainly distributed in low latitudes, in particular in the southern hemisphere. In the context of foreland structures, carbonate rocks mainly developed in the south of England, the Hercynian Rift Trough, the northern slope of Alaska, Verkhoyansk, and northern Siberia, whereas mixed carbonate and siliceous clastic rocks accumulated in Kazakhstania, the South China continent, Southern Europe, and North Africa [4]. In the Early Carboniferous, shallow-water carbonate rocks developed widely, but oil and gas reservoirs mainly developed in the North American cratons and Ural foreland depression belts. The rapid sea level rise in the Early Carboniferous resulted in extensive transgression in the craton shelf areas and high abundance source rocks developed there. In addition, extensive molasses and continental river-lake clastic rock formations developed in these foreland basins. The size of oil and gas reservoirs often reflected the varying control effects of the foreland structure environment.

4. Upper Carboniferous–Lower Permian Series

During Upper Carboniferous–Lower Permian Period, the convergence and collision of continents led to the formation of the Pangea supercontinent. West Gondwana and Laurasia collided and spliced. The strong compression caused large-scale thrust and magmatic activity along the east and south of North America. With the expansion of the Paleo-Tethys Ocean and the closure of the Rick Ocean, the Variscides Fold Belt and Molasse Basin developed in Europe. At the same time, craton landmasses in Kazakhstania and Siberia collided and sutured with the Eastern European plate along the Urals [5]. In the early Permian, cratons in North China collided with the Tarim landmass and Laurasia, while the Iran, Qiangtang, Malaysia, and Indonesia landmasses split from Australia. The New Tethys Ocean began to enlarge and the Paleo-Tethys Ocean began to close. A series of foreland basins developed along the Volga–Ural, Hercynian, and Kazakhstania–Siberian Fold Belts. The Upper Carboniferous–Lower Permian Series included five secondary super-sequences.

During the Upper Carboniferous–Lower Permian Series, reservoirs all over the world included shallow-water granular and phyllophyte bio-built limestone, the quality of which depended on the Paleo climate. Humid weather conditions led to selective dissolution of unstable carbonate minerals, leading to the development of secondary pores. Siliceous clastic rock formations were also well developed in the Pennsylvania and Early Permian, which included the development of shallow seas, wind-formed and eroded valley filling, deltas, and basin floor fan deposits. For example, abundant surface sea deposits developed on the North China plate, leading to development of large-scale coal rocks.

During the Upper Carboniferous–the Lower Permian Series, larger mosaic-type carbonate platforms often developed along the margins of cratons, whereas smaller mosaic-type platforms and isolated platforms mainly developed in the interiors of the craton basins. In addition, saprolite mud shale, siliceous clastic rocks, and evaporite generally developed in craton basins. Most of the known carbonate reservoirs were located in central and southwestern United States, the Volga–Ural area, and southern China. The active structural environment was conducive to the development of structural or structural-stratigraphic composite reservoir traps [6].

5. Middle-Upper Permian Series

During the Middle-Upper Permian Series, tectonic activities all over the world were relatively weakened. Foreland basins that had developed along the convergent and collisional boundaries created by the joining of the Carboniferous and Early Permian Pangea supercontinents began to fill up during this period. The ocean–land collision around the eastern Pacific led to the development of a series of retroarc foreland basins along the front edge of Andes in the western part of the Northern America continent. The Paleo-Tethys Ocean subducted northward and reduced in Eurasia, resulting in a series of fore-deep depressions. With the separation in the Early Permian of the Kilimian continent and the China landmass from the Australian, Indian, Arabian, and African plates along the north or northeast, the New Tethys Ocean began to develop, forming the Arabian Inland Shelf Basin. The Triassic can be divided into two second-order supersequences, namely, the collisional assembly of the Pangea continent (the Pennsylvanian or the Early Permian) and post-collision tectonic events (the Middle-Late Permian). In the Middle Permian, the ice chamber climate that prevailed in the Late Carboniferous began to weaken and quickly transitioned to a greenhouse climate. At that time, the orogenic belts resulting from splicing of the pan-continental shelf had a great effect on the global climate. Therefore, the continental deposits and the marginal sea deposits during the Permian were composed of red beds, eolian rocks, tidal dolomites, and evaporites, which explains the widespread arid and semiarid climatic conditions all over the world. For example, the Hercynian/Appalachian Mountains traversing the equator blocked the eastward trade winds from the subtropics, which resulted in a tropical rainy climate. Therefore, in the Permian, most carbonate platforms developed in marginal or inland sea basins.

In the Permian, the mixed sedimentary system of carbonate platforms and carbonate-siliceous clastic rocks mainly developed at the 50° north–south latitude, especially in the Tethys region. Specific carbonate platform development areas included the Barents Sea, North West Greenland, the northern slope of Alaska, the Volga–Ural area, the ancient continental margin basin in the west of North America, the Iranian Kirimi landmass, Qiangtang, Malaysia, Indonesia, Indochina, and the South China Micro-terrain. Controlled by features of the previous foreland, tower reefs and line reefs developed on the narrow foreland edges. At the same time, the orogenic belts created by the splicing of the Pangea continent controlled the global climate and the subsidence and deposition of the post-orogenic basins. As a result, basins located in the tropics became very limited, leading to the development of thick-layered evaporites, large-scale dolomites, and subsequent karstification [7]. In the Late Permian, the carbonate deposition system gradually decreased, whereas large-scale clastic rock formations gradually developed.

6. Triassic–Lower Jurassic Series

In the Early Triassic, the Pangea supercontinent began to disintegrate. Though continental separation did not occur until the Jurassic Period, many intracontinental basins and aulacogens developed. With the gradual expansion of the New Tethys Ocean, the Ancient Tethys Ocean shrank, and the collision and joining of the South China and North China continents occurred. In the Early Jurassic, a series of collision events occurred along the northern part of the New Tethys Ocean, resulting in the closure of the Paleo-Tethys Ocean. At the same time, the New Tethys Ocean transitioned from central expansion to a rift in the south. Large-scale transgressions proceeded to many inland aulacogens and rift basins.

The Triassic and Early Jurassic form a first-order mega stratigraphic sequence corresponding to a relatively symmetrical first-order sea-level rise and fall cycle. It included 2 second-order sequences and 18 third-order sequences (12 in the Triassic and 6 in the Lower Jurassic Series). Following the ice chamber climate in late Palaeozoic, greenhouse climate gradually prevailed in the Triassic.

During the Lower-Middle Triassic Series, carbonate platforms developed in the western margin of the New Tethys Ocean and the Paleo-Tethys Ocean (Arabia, south-central Europe, and the area between the Caspian Sea and the Black Sea) and the eastern margin of the Paleo-Tethys Ocean (South China, Indochina, Malaysia, and Indonesia); mixed carbonate and clastic rocks were deposited in isolated and united continental fragments in western Canada and North America. During the Lower-Middle Triassic Series, carbonate reservoirs were distributed in the Arabian Inland Shelf Basin, with isolated platforms in southern Europe and drifting and discrete Tethys landmasses. While during the Late Triassic, carbonate platforms developed mainly on the western and southern margins of the Neo-Tethys, whereas mixed formations of carbonate and clastic rocks developed on the southern margin of the Neo-Tethys and western and northern Canada [8]. At this time, large-scale clastic rock deposits replaced carbonate platforms on the China landmass. In the Early Jurassic, affected by the extinction event at the end of the Triassic, the lack of framework clastics and the enrichment of shale limited the development of carbonate rocks, but argillaceous facies developed in local areas, providing good source rocks. At the same time, between North America, South America, and Africa, internal landing rifts and aulacogens developed. Clastic rocks were accumulated in large-scale rivers, lakes, and deltas.

7. Middle Jurassic–Lower Cretaceous Series

During the Jurassic Period, new ocean floor expansion occurred in the Atlantic Ocean and Gulf of Mexico areas, and intracontinental rift between India, southwest Australia, and the Antarctic Continent also became active. At the same time, accompanied by a series of complex rift valley and seafloor expansions along North Africa, many small plots were produced in the western "Bay of Tethys." The ancient Tethys Ocean closed up in the Middle Jurassic [9]. In addition, the Larami Orogenic Belt along the front edge of the Cordillera Volcanic Arc also began to develop. In the Late Jurassic and Early Cretaceous, against the background of the global tectonic disintegration of supercontinents and the drift of the continental blocks, the global plates began to readjust.

The Middle Jurassic included one first-order sea level rise-and-fall cycle, and four second-order sea level rise-and-fall cycles that were globally comparable to a certain extent. The sea level decline in the Late Jurassic led to the development of large-scale clastic rocks. At the same time, evaporites accumulated on a large scale, especially in the Circum-Tethys. Strongly affected by the magmatic activities in Antarctica and Namibia, greenhouse climate prevailed during the Mid-Late Jurassic and the climate gradually became cooler and arid during the Late Jurassic–Early Cretaceous. In fact, climatic zoning was very obvious in the Cretaceous, which was the reason why coal and rock were enriched in high-latitude areas.

In the Middle Jurassic, at the narrow rifted continental margin (including the Atlantic Ocean, the southern Gulf of Mexico, and the Neo-Tethys block) and the broad passive continental margin (including India, Madagascar, Arabia, and the "West Tethys Bay"), combined carbonate platforms developed. At the same time, carbonate platforms also developed in the west of South America, Qiangtang, northwestern Tethys, and Indonesia. In the Late Jurassic, carbonate platforms developed in major passive continental margin environments such as the Gulf of Mexico, South America, North Africa, and the southern margin of Tethys, whereas isolated platforms controlled by structure developed in the northwest of Tethys. In the Jurassic, the humid climate at high latitudes enriched coal and rocks. In the Late Jurassic, the sea level dropped and the hot climate also caused evaporite to accumulate on a large scale.

8. Cretaceous System

During the Cretaceous, the Pangea supercontinent disintegrated, and intense expansion and plate drifting of Haiti occurred. With the expansion of the Atlantic Ocean, the African continent was separated from the South American continent when the European continent and Greenland were torn apart from the North American continent [10]. In the Early Cretaceous, the separation of the Indian landmass, the Australian landmass, and the African landmass from the Antarctic continent accelerated the closure of the New Tethys Ocean.

First-order sea level rise continued all over the world to during the Cretaceous and reached the maximum surface flooding during the Turonian Period. The Cretaceous included three main second-order sequences. The expansion of the seafloor and the intensification of volcanic activity, along with continuous transgression, strengthened the greenhouse climate of the Triassic and Jurassic Periods, which intensified during the Cretaceous.

Carbonate rocks from the Early-Middle Cretaceous were distributed mainly in the Gulf of Mexico, the northwestern and northern shelf of South America, the periphery of the New Tethys Ocean, the northern edge of the Tethys Ocean, the Arabian platforms, and local parts of the continental shelf on both sides of the Atlantic Ocean. In addition, small carbonate rock platforms developed in the passive continental margins of Australia, New Guinea, and northern India. With the continuous rise of sea level, small carbonate platforms developed at some of the abovementioned areas during the Late Cretaceous. At the Arabian platform, the northern edge of India, and the southern edge of Tethys Bay (including the mid-sea area, the southern part of Eurasia, and the edge of the North African continent), carbonate rocks developed toward the land. At the Gulf of Mexico, the northern margin of Tethys Ocean, and the Atlantic coast, carbonate platforms retreated or were submerged. At this time, pelagic deposits generally accumulated on the northwestern coast of Europe, the North Sea, the northern margin of the Gulf of Mexico, and local parts of the west coast of North America. In fact, the development and distribution characteristics of carbonate rocks depended on the regional or local tectonic environment. For example, the Laramie Orogenic Movement controlled the development of large oil fields in the Golden Lane area to a certain extent. At the same time, affected by the combined effects of the tensile tectonic environment, hot paleo-climate, and continuous global sea level rise, clastic rocks developed in the Cretaceous, the distribution of which was complicated.

9. Tertiary System

In the Tertiary system, the active rifting and dispersion of the Mesozoic gradually transitioned to agglomeration and evolved toward the next supercontinent. In the context of compressional island arcs and foreland, many extensional basins, such as back-arc foreland basins, were formed in local areas. In these basins, lacustrine sedimentary systems spanning large areas were widely distributed; Palaeogene deep–semi-deep lacustrine argillaceous facies and coal source rocks often filled graben and half-graben depressions. In addition, the stretching and dispersion that occurred during the Cretaceous Period continued to develop. For example, the Mid-South Atlantic Ocean continued to expand and passive continental margins developed. In the Red Sea area, rifting was episodic during the Palaeogene. Starting from the Neogene, the Red Sea gradually became a basin [11].

After the largest flooding in the Late Cretaceous, the first-order sea level all over the world began to fall during the Tertiary. The Tertiary Series included five main second-order sequences. Greenhouse climate continued all over the world during the Cretaceous and reached its peak in the Early Palaeogene. However, with the gradual decline of the sea level and the accumulation and uplift of continents, the Neogene was characterized by an icy climate.

Palaeogene carbonate platforms were mainly distributed in the Yucatan Peninsula and the Gulf of Mexico, the coast of North Africa, the Arabian Peninsula, and the Florida Peninsula. On the coast of West Africa, South Africa, and Madagascar, narrow and open shelf carbonate deposits developed. Considerable carbonate rocks developed in Papua New Guinea and along the passive continental margins of northwestern Australia. In addition, some small isolated carbonate platforms developed in the forelands of Southern Europe and the Central Sea. In the Neogene, joint inlaid carbonate platforms often developed on structurally active land blocks [12]. Open continental shelf carbonate deposits were mainly distributed in the passive continental margin of northwest and north Australia and the Caribbean, while gentle slope platforms developed on the southern coast of Australia. In addition, isolated reef platforms emerged in the active continental margins of Southeast Asia.

At the same time, the discrete-converging transitions all over the world and the transitional climate conditions of the temperate–glacial period served as a backdrop for the development of a wide array of geographical features: large-scale continental rivers, lakes, deltas, deep-sea facies, and gravity flow and other clastic rock formations. In summary, the characteristics of these depositional elements and lithofacies are then complexity and variability.

References

[1]   Jianghai Li, Hongfu Jiang. *Reconstruction Lithofacies Paleogeography and Environment Atlas of Global Ancient Blocks*. Beijing: Geological Publishing House, 2013.

[2]   Deliang Liu. *Discussion on Hebei-Shandong Meridional Tectonic Zones*. Beijing: Geological Publishing House, 1989.

[3]   Guangding Liu. Geophysical field and geodynamic characteristics of China Sea. *Acta Geology*, 1992 66(4): 10–24.

[4]   Kewen Gan, Jianyi Hu. *Atlas of World Petroleum Basins*. Beijing: Petroleum Industry Press, 1992.

[5]   Zengmiao Guan. *Oil and Gas Resources and Exploration in African*. Beijing: Petroleum Industry Press, 2007.

[6]   Yuzhu Kang. Discussion on oil and gas distribution law and oil prospecting direction in Tarim Basin. *Earth Science*, 1991 16(4): 79–86.

[7]   Yuzhu Kang, Zhijiang Kang. Significant progress of geomechanics in oil and gas exploration in Tarim Basin. *Chinese Journal of Geomechanics*, 1995 1(2): 1–10.

[8]   Yuzhu Kang. *Paleozoic Marine Oil-forming Features in China*. Urumqi: Xinjiang Science and Medical Publishing House, 1995.

[9]   Yuzhu Kang. *Main Tectonic Systems and Oil and Gas Distribution in China*. Urumqi: Xinjiang Science and Medical Publishing House, 1999.

[10] Paul J. J. Welfens, David, B. Audretsch, John T. Addison. *The Global Economy: R&D, Structural Change and Employment Shifts*. Berlin, Heidelberg: Springer, 1998.

[11] Werner, R., Valev, D., Danov, D., *et al.* Study of structural break points in global and hemispheric temperature series by piecewise regression. *Advances in Space Research*, 2015 56(11): 2323–2334.

[12] Youjin, S., Jiazheng, Q. Strong earthquake activity and its relation to regional neotectonic movement in Sichuan-Yunnan Region. *Earthquake Research in China*, 2001 15(3): 239–251.

Chapter 3

Features of Crustal Stress All Over the World

Abstract

This chapter discusses characteristics of global geostresses, including detailed descriptions of global geostress as well as that in China, the analysis of the relationship between crustal geostresses and earthquakes, and the analysis of causes and consequences of earthquakes.

Keywords: Global, crust, geostresses, earthquakes.

Nowadays, information regarding the magnitude and direction of crustal stress all over the world can be organized into these key maps: the global level change map, the global gravity map, and the global heat flow map. An integration of the stress and motion data of all continents would provide a better understanding and simulation of the dynamic processes driving tectonic movements.

1. Crustal Stress State All Over the World

In 1975, Panall and Chanden published the first global stress direction diagram, which included 59 stress relief measurements. In 1979, Richardson *et al.* published a world stress map with data from approximately 133 points in total, which included mainly stress derived from the focal mechanism and some additional data from the surface core. It has been the aim of the International Lithosphere Program (ILP) since 1986 to put together all available and expected data reflecting the tectonic stress and expressing them on plane figures as per relative magnitude and direction of the principal stress, which has been implemented through the Second Five-Year Plan as part of the World Stress Map project [1]. Dr. Mary Lou Zulack from the United States acted as Chair of the project, and more than 30 scientists from 16 countries participated in it.

The Global Stress Map project collected 7,328 *in-situ* points of stress direction data, of which 4,413 points had reliable tectonic stress signs with deviation recorded along the horizontal stress direction at $<\pm25°$ [2]. The *in-situ* stress measurement results were in good alignment with the geological observation results for the upper 1–2 km. Data of inclined boreholes can reflect the stress state at a depth of 1–4 km, as well as at a depth of 5–6 km in some cases. Results obtained from the focal mechanism solution can be applied to a depth of approximately 20 km. The thickness of the upper brittle lithosphere is approximately 20–25 km. The large-scale uniform area stress field within the range of 20–200 times this thickness is named the first-order stress field. Typical wavelength range of the second-order stress image is more than 5–10 times [3] the thickness of the upper brittle lithosphere. This chapter mainly discusses the first-order stress field.

Zoback [2, 4] rounded up points based on the world stress diagram showing the first-order stress state (Table 1):

(1) At most parts all over the world, there is a uniform stress field at the entire upper brittle crust.
(2) The central part of most continents is dominated by compression stress (thrust and strike-slip state) and the maximum principal stress is horizontal.
(3) The high-altitude areas of continents and oceans are tensile stress areas (normal fault motion stress states), where the maximum principal stress is generally vertical.
(4) The regional consistency of direction and relative magnitude of stress can be used to determine the large-scale zoning of regional stress, which is mostly consistent with geological regions, in particular tectonic active regions.

There were regions with a uniform S_{Hmax} direction, such as the eastern part of North America, the western Canadian Basin, central California, the

Table 1.   Global first-order stress relief (according to Zoback [2]).

Areas	S_{Hmax} or S_{Hain} direction	Stress state
North American Plate		
Central region of the plate	ENE	T/SS
West Cordillera	The stress state is too complicated and goes beyond the scope of discussion.	
Central America and Alaska		
South American Plate		T/SS
Mainland regions	E	
High altitude in the Andes	N	NF
Eurasian Plate		
Western Europe	NW	SS
China/East Asia	N-E	SS
Qinghai–Xizang Plateau	WNE	NF
African Plate		
East African Rift Valley	NW	NF
Middle of the plate (West Africa and South Africa)	E	SS
North Africa	N-NW	T/SS
India–Australia Plate		
India	N-NE	T/SS
Central Indian Ocean	N-NW	T/SS
West Indian Ocean	N-NW	NF
Central Australia and northwest shelf	N-NE	TF
Coasts along Australia South Sea	E	TF
Pacific Plate		
Younger crust (<70 km)	NE	SS
Older crust (>70 km)	NW?	T/SS
Antarctica plate		
Central plate	?	?
West Antarctic Rift Valley	E-NE	NF

Notes: NF — normal fault motion stress state; SS — strike-slip fault motion stress state; TF — thrust fault motion stress state; and T/SS — thrust and strike-slip motion stress state.

Andes Mountains, Western Europe, the Aegean Sea, and Northeast China. The stress orientations at major regions of plates all over the world are listed in Table 1.

The first-order stress details provided in Table 1 state the relative importance of various large-scale stress sources acting on the lithosphere:

(1) The direction of the compressive stress field in the middle of the continent is mostly the result of the compressive stress (caused by ocean ridge compression and continental collision) exerted on the continental boundary and is mostly controlled by the shape of the continental boundary.
(2) The horizontal tensile stress caused by buoyancy in high-altitude areas is a second-order stress field, which often relates to special geological or structural features and has a disturbance or local control effect on the compression of the central part of the continent caused by continental boundary forces.
(3) It is hard to estimate the effect of drag force by stress direction data alone.

In the middle of some continents, there was a clear correlation between the S_{Hmax} direction and the absolute motion direction of continents. There was a significant positive correlation between the S_{Hmax} direction and the absolute motion direction of the middle of the North American continent (including most of the central and eastern parts of the United States, most of Canada, and possibly the western Atlantic Basin) and the South American continent, the two continents with the fastest movement. In Western Europe, the relationship between the S_{Hmax} direction and the absolute direction of motion was ideal, with the exception of the Aegean Sea. This correlation was manifested as the clockwise rotation of the S_{Hmax} direction relative to the absolute direction of motion, that is, relative to the absolute velocity field in the direction of WNW. The observed S_{Hmax} direction is more northerly [5]. However, the relationship between the S_{Hmax} direction and absolute motion direction of continents within the Pacific continent was less obvious. The stress image of eastern Asia was strongly affected by the collision between Eurasia and the Indian–Australian continents. Within the Himalayan collision zone, the S_{Hmax} direction was generally along the NS direction, which was almost

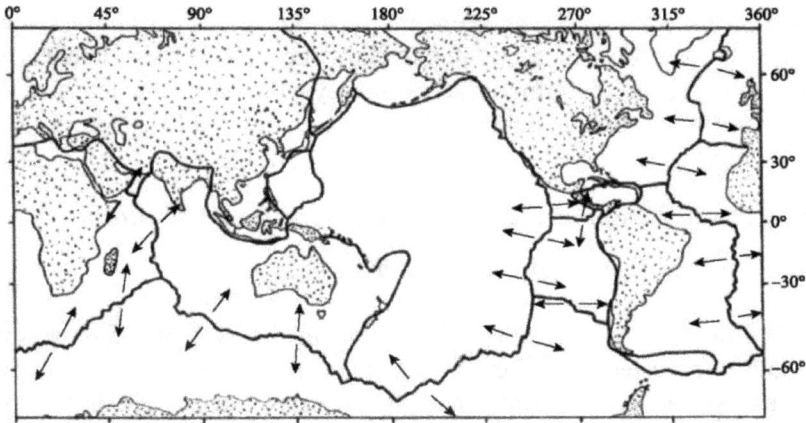

Figure 1.   Directions of ocean ridges and ocean ridge thrust (according to Richardson [6]).

parallel to the absolute movement direction of Eurasia and the relative (convergence) movement direction between Eurasia and India–Australia [7]. However, in eastern China, the trajectory of $S_{H\max}$ formed a quasi-radiation image, which was oblique with the absolute and relative motion directions of the continent (Zoback, 1989).

Richardson [6] used ocean ridge thrust (Figure 1) to explain the consistency between intracontinental stress and the absolute motion direction of continents, and demonstrated the main role of ocean ridge thrust in intracontinental deformations, in particular, in the formation of uniform stress directions in a large area. The comparison between oceanic ridge torque and the absolute motion direction of a continent showed that there was an extremely obvious correlation between these two directions in the following continents: the Pacific Ocean, Cocos (Keeling) Islands, North and South America, Arabia, and India–Australia.

2. Crustal Stress States of China

Research on the crustal stress state in China, especially *in-situ* stress measurements, was developed in the early 1960s under the advocacy of Mr. Siguang Li, and has been widely applied by departments relating to earthquakes, geology, metallurgy, coal, petroleum, and hydropower.

Based on *in-situ* stress measurements, focal mechanism solutions, topographic deformation measurements, fault micro-displacement measurements, and seismic deformation zone surveys, the basic characteristics of the crustal stress state in China have been summarized [8].

2.1. *Crustal Stress in the Horizontal Direction Dominates*

The following are the main bodies of evidences:

(1) The amount of horizontal dislocation in a series of major seismic faults generated since the 1920s was generally greater than their vertical dislocation and it was especially the case with seismic faults of the 1920 Haiyuan Earthquake, the 1931 Fuyun Earthquake, and the 1973 Luhuo Earthquake, the horizontal offset of which was about 5–12 times the vertical offset [9].
(2) Trace displacement measurement results of faults reflected that the horizontal activity of some major active faults in the western part of the Chinese continents and North China was relatively large, which was several times their vertical activity.
(3) Results of *in-situ* stress measurement also showed that there were strong horizontal stresses in the crust of China, especially the stresses controlling the large earthquake activities in the southwest, northwest, and north China, where the horizontal component was dominant.
(4) Results of focal mechanism solution proved that most of the focal sources in China were mainly horizontal displacement and that the direction of the maximum principal compressive stress was almost horizontal.

2.2. *Crustal Stress Varies with Depth*

The following are current variation features of crustal stress activity along with depth in China [10]:

(1) The maximum principal stress and shear stress values increase with depth, whereas the change rate varies greatly depending on the area or structure.

(2) The ratio of the average horizontal principal stress to the vertical principal stress decreases as the depth increases.
(3) The direction of the maximum principal stress does not change much along with depth.

2.3. *The Current Crustal Stress State Shows Zonal Features*

The current crustal stress state in China shows obvious zoning features, which can be divided as follows:

(1) The western regions, including Xinjiang, Xizang, and the western part of Gansu, Qinghai, and Sichuan, were areas continuously suffering from near south–north compressive stress.
(2) The eastern regions include the Northeast, North China, South China, and Central South. At present, the tectonic stress state is generally close to the east–west direction. However, there is still a sub-level zoning. The direction of the maximum principal stress in North China–Northeast China is near east–west from north to east, while in Southern China, the NW direction dominates.
(3) The eastern parts include Gansu, Qinghai, Sichuan, and Yunnan. It was the transitional area between the east and west. Results of *in-situ* stress measurement showed that the direction of the main compressive stress was nearly east–west. The baseline survey data of the seismic deformation zoning and the crossing fault reflected that the activity mode of the fault in this area was east–west compression [11]. The P-axis azimuth for earthquakes of magnitude 6 or above was also close to east–west.

The intensity of current crustal stress activity in eastern and western China was also significantly different. The intensity of stress activity in the west was high, while it was low in the east.

The pattern of the current crustal stress state in China is basically a continuation of the tectonic stress field since the Neogene. Interaction of plates surrounding China had a significant impact on China's stress field. However, it is still a matter of controversy which plate influences China's stress field and what role it plays on the boundary of each plate, whether squeezing or pulling. Zhenliang Shi *et al.* [12], Qidong Deng *et al.* [13], and Ogata *et al.* [14] analyzed the distribution of earthquakes, seismic fault plane solutions, and geological data, maintaining that the compression of the Indian Plate, the Pacific continent, and the Philippine Sea continent on the Chinese Mainland was the main source of the Chinese stress field. Tapponnier and Molnar [15] pointed out after studying the tectonic movement in China and Southeast Asia that the main source of power was the collision between the Indian continent and Eurasia and that the interaction between the Pacific Continent, the Philippine Sea, and Eurasia had little effect on China's tectonic movement. It has also been pointed out that the stress field in East Asia was mainly caused by mantle convection. The simulation results of the stress field in China showed that though both the continental boundary force and the drag force formed by the mantle convection played a role, the boundary force played a leading role.

Recently, Shaoxian Zang *et al.* [16, 17] discussed this issue by combining seismic data and geological data, and then put forward the following idea: the Pacific Ocean dove under Eurasia along the bottom of the Japanese Trench, which was well coupled with Eurasia, and produced a squeezing effect in Northeast China. The compression direction was approximately N85°W near the Japan Trench and the main area where it had an impact on the stress field in China was located between 35°N and 42.5°N. The Philippine Sea continent dove under the Ryukyu Island arc. As these two continents were not well coupled, and the Okinawa Trough was expanded, no squeeze effect was produced on North China. The Philippine Sea continent collided with the Eurasia at around 121.8°E between 21.5°N and 24.2°N, forming a strong NWW squeezing effect on Taiwan province and the southeast coast of China, but the impact was limited. From places south of 22.9°N, the South China Sea dove under the Philippine Sea, so that the Philippine Sea continent did not squeeze the South China Sea [18]. The Indian continent collided with Eurasia in the Himalayan region, forming the NNE-directed principal compressive stress field, which was the main source of power for the stress field in China. The west wing of the collision zone may have been a new convergence zone, with the main compressive stress axis perpendicular to the direction of the seismic zone. The east wing of the two plate boundaries suddenly turned at 26.5°N and 97°E, forming the Burma mountain arc subduction zone between 20°N and 26.5°N and causing a compression effect along the NE-NNE direction. But such compression existed in the west of the Hengduan Mountains, and the primary compressive stress direction to the east of the Hengduan Mountains was SSE-NNW [19]. The

Indian continent also formed the NW and NNW principal compressive stress fields in the southeast of the Chinese continents and the South China Sea, and the direction of the pressure axis further south was closer to the NS direction.

The characteristics of the tectonic stress field of the Chinese continents as inferred by Zhonghuai Xu *et al.* [20] based on data of 9,621 *p*-wave initial motion directions of 5,054 small earthquakes ($1 \leq ML < 5$) were basically similar to those inferred by Shaoxian Zang *et al.*, but were more in harmony with the world stress diagram (Figure 2). The horizontal direction dominated in the largest and smallest principal compressive stress axes in the entire Chinese continents, showing that the seismic tectonic deformation was mainly caused by strike-slip fault activity. The principal stress direction presented a fairly regular radial image, that is, the maximum horizontal compressive stress trajectory radiated from the Qinghai–Xizang Plateau to the north, east, and southeast of the Chinese continents, while the minimum horizontal compressive stress direction was located on the arc protruding outward from the Qinghai–Xizang Plateau. This overall picture shows that the stress field controlling tectonic activity and earthquakes in the Chinese continents was not mainly stimulated by internal local causes but was closely related to the movement of the surrounding plates. The exception is that in some areas of Northern China, such as the southern section of the Fenwei Graben, northern Jiangsu, and northwestern Liaoning, the maximum principal compressive stress axis was upright.

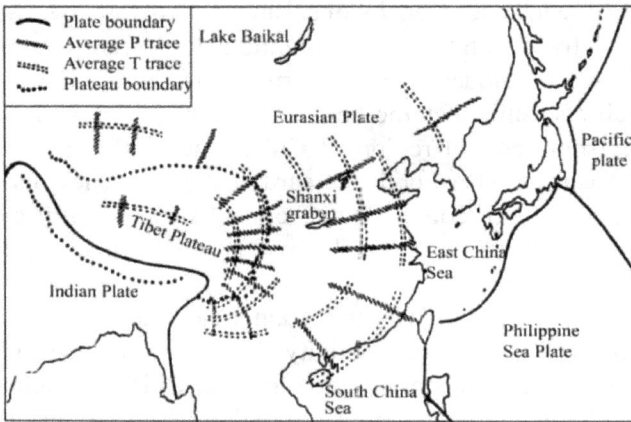

Figure 2.   Maximum and minimum horizontal stress trajectories inferred from focal mechanism data (according to Zhonghuai Xu [20]).

In some areas in the west, such as the northeastern edge of the Qinghai–Xizang Plateau and the Tianshan Mountains, the minimum principal compressive stress axis was upright. At the eastern end of the Himalayan Arc, the average T-axis trajectory converged in an arc to the south of Assam. It indicated that the India–Burma arc boundary played an important role in controlling the southwestern stress field in China.

3. Relationship between Crustal Stress State and Earthquakes

The earthquake process is a mechanical process. The occurrence, distribution, activity level, rupture mode, and characteristics of the macro-seismic deformation zoning of earthquakes are closely related to different crustal stress states [21]. However, such relationships are extremely intricate and are learned only based on experience so far. The work of predicting earthquakes by changes of ground stress measurements is being further explored at home and abroad, and yet no reliable criterion for prediction is available.

Results of seismic activity and *in-situ* stress measurement showed that large earthquakes either in China or all over the world occurred mainly in tectonic active areas where horizontal or near-horizontal tectonic stresses dominated. In contrast, in areas where horizontal stress is equivalent to vertical stress or vertical stress is greater than horizontal stress, the frequency and intensity of seismic activity are low. With regard to the stress activities on the active tectonic belt of occurrence and distribution of large earthquakes in western China, North China, and the Circum-Pacific seismic tectonic belt, the horizontal component is dominant. In the south-central region of China, the Ural region of Russia, India, and Australia where both the earthquake frequency and intensity are low, the horizontal component of crustal stress is very close to its vertical component, or the vertical stress is greater than the horizontal stress [22, 23].

The level of seismic activity in certain areas is also closely related to the intensity of tectonic stress activity there. The intensity of tectonic stress activity in the western region of the Chinese continents was 5~6 times higher than that in the eastern region. Thus, the seismic activity level in the western region was significantly higher than that in the eastern region. The strong and weak features of seismic activity in some other earthquake-prone areas in the Chinese continents were also specific representations of changes in the intensity of tectonic stress activity in the corresponding area over time [24].

Observations and analyses of the geostress state of focal points before and after the Tangshan earthquake and surrounding regions showed that geostress changed significantly before and after the earthquake. During the period from 1968 to 1971 before the Tangshan earthquake, initial movement signs, the small earthquake P-wave, at the Changli and Douhe Stations showed an irregular and scattered distribution on the projection sphere, indicating that the Tangshan–Luanxian area did not show a certain strengthening orientation of dominant stress during this period. It can be considered that it was under long-term stable stress. From 1972 to June 1976, the stress was predominantly distributed and a unified force field was formed, which was dominated by horizontal stress and was manifested as being squeezed from east to west and tensioned from north to south. With the energy of the main earthquakes and strong aftershocks being released, the superiority of the stress field in this area was significantly weakened. Before the Tangshan earthquake, tension jumps perpendicular to the fault direction were observed at the Douhe and Zhaogezhuang Stations near the main fault of the earthquake. According to the leveling observation data across the fault before the earthquake, it was believed that it was the reflection of the tensile crack in this direction near the surface, which was caused by local protrusion of the crust along the fault.

4. Causes and Consequences of Earthquakes

Mr. Siguang Li once said, "The critical point is that the shaking is inevitable due to a sudden rupture of the underground rock at a certain point. The rupture must have been caused by a force that continued to build until it exceeded the strength of the rock's resistance. The reason why a rock bursts is that there must be a constantly strengthening force (mechanical force), which exceeds the confrontation strength of the rock. The strengthening of that force must have a process of accumulation. The problem lies in the process of accumulation."

There are approximately 50,000 earthquakes felt in a year. The aftershocks of a major earthquake range from dozens to hundreds of times. According to observations, the subsidence process of the alluvial plain in

Hebei, China, is too slow, with an average subsidence of approximately 1mm per year. However, a strong earthquake occurred in Xingtai, Hebei, in 1966. Results of measurements by the National Surveying and Mapping Administration showed that the fault zone produced by the earthquake matched with the range of the extreme earthquake zone. The NNE-trending fault zone produced by the earthquake had a subsidence range of 315–714 mm, whereas the maximum increase in the neighboring area was only 40–72 mm. In the early morning of July 28, 1976, another 7.8-magnitude earthquake occurred in Tangshan, Hebei, and its subsidence was even greater. The area almost fell to ruins. In 1885, a sea earthquake occurred in the Adriatic Sea near Italy, which caused a change in the depth of the seabed from 200 to 3,000 m. During the eruption of Mount Pelée on Martinique Island, the depth of the neighboring parts of the sea increased by several hundred meters. After three or four volcanic eruptions, the Japanese island of Krakatau sank into the sea. From analysis of these specific examples, it has been determined that rocks at depths with strong geostress were under increasing pressure when a fracture suddenly occurred and the earthquake occurred when it went beyond the limit of compressive strength of the rocks. That is only natural.

Likewise [25], large oceanic basins of enormous size subside as the center of gravity of subsidence breaks down at the site. At a later stage, the deeper or abyssal rock basement, from being initially a tiny opening of the lower part of the rock, develops a large seaquake with the sudden major rupture of a robust subsidence process. In this manner, heat, gas, liquid, and basalt magma of the "asthenosphere" below the lithosphere invaded along the fault line on a large scale using the upward force and gushed out of the bottom of the ocean. As the pressure difference of the large tensile ruptures was small, no substance was ejected into the air, showing gentle characteristics. The magmatic belt of the fault activity evolved from the original settlement stage and gradually returned into the original stage. As the force evolved, magma activity was first dominated by basic activities, then by neutral ones, and later by acidic ones. The scale of magma intrusion and eruption gradually increased, forming a majestic mountain system and causing subsidence in many places in the adjacent ocean basins, deep horizontal planes, and continental coasts. The proportion of areas that became deep ocean and deep sea was increasing. For example, ocean basins, deepwater plains, and sea bays were the main subsidence areas on the earth's surface, while continental basins, lakes and swamps, alluvial plains, and downstream deltas were the slow secondary deep subsidence areas. It is worth noting that sinking places on the earth occupied a large portion of the earth in ancient times, approximately three quarters of the earth.

Ground stress of earthquakes occurring in plateaus and high mountain areas produced strong squeezing, a certain degree of horizontal displacement, and ascending and thrusting action on rising areas. For example, on May 12, 2008, an 8.0-magnitude earthquake occurred in Wenchuan, China, which was the result of compression stress of the Indian landmass on the Asian landmass. But the landmasses did not separate.

When an earthquake occurred in 1970 in Yunnan Province, a 60-km-long northwest-trending fault experienced significant horizontal displacement. The maximum horizontal displacement seen in the field was 2.2 m, but the vertical displacement was only tens of centimeters. Another example was the 8.1-magnitude earthquake that occurred in Nepal on April 25, 2015, the focal depth of which was 15 km underground. Due to the northward squeezing of the ocean basin of the Indian Ocean and the Bay of Bengal and Arabian Sea along the coast, Nepal and the Himalayas were strongly squeezed, with uplift and thrust movements. In this major earthquake, a crust, approximately 120 km long and 60 km wide, moved about 3 m to the south and the angle between the fault and the ground was only 10°, which caused the overall crust in the area to shrink. However, it is worth noting that the cracks were generally in the suspended parts of plateaus and alpine regions, and cracks were almost in a tensile fracture state, and were big at the top and small at the bottom [26]. When it came to the later period, due to the strong action of ascending and horizontal movements, tensional rift zoning with large suspension was produced, which under the effect of earthquakes, magma intrusion, and volcanic eruption can gradually transform into inland seas belonging to subsidence areas, such as the Red Sea and the Mediterranean Sea, and could further develop toward the direction of oceans.

The major landmasses will not move. The earth's lithosphere can be divided into six continents. These land masses are above the "asthenosphere" and are an integral whole. They have never separated, are not separated and will never be separated. The reason six landmasses are observed now is that oceans have separated them and they are still a complete crust when oceans are removed.

As such, the author believes that there was no continental drift and plate movement at all and that it was the crustal fault activity and subsidence after orogeny in different directions and of different natures, because of squeezing, tension, compression, and torsion produced locally as a result of crustal tectonic movement, that changed the scope and direction of the oceans.

References

[1] Zhijun Jin, Jinyin Yin. *Features of Petroleum Geology and Oil and Gas Distribution in Asia.* Beijing: China Petrochemical Press, 1997.

[2] Zoback, M. L. First- and second-order patterns of stress in the lithosphere: The world stress map project. *Journal of Geophysical Research: Solid Earth,* 1992 97(B8): 11703–11728.

[3] Guoyu Li, Zhijun Jin. *Atlas of World Petroleum Basins.* Beijing: Petroleum Industry Press, 2005.

[4] Zoback, M. L., Zoback, M. D. Tectonic stress field of the continental United States. *Memoir of the Geological Society of America,* 1989 172(1): 23–37.

[5] Luofu Liu, Yixiu Zhu. *Petroleum Geological Features of Coastal Caspian Basins and Central Asia.* Beijing: China Petrochemical Press, 2007.

[6] Richardson, R. M. Ridge forces, absolute plate motions, and the intraplate stress field. *Journal of Geophysical Research,* 1992 97(B8): 11739–11748.

[7] Jichen Liu. Continental block tectonics of North Qilian orogenic zone. *Earth Science,* 1991 (6): 39–46.

[8] Qiusheng Zeng. The present state of crustal stress in China. *Journal of Geomechanics,* 1989 1: 11.

[9] Shenshu Liu, Jinhai Zhao, Beiyu Xiang. *Structure and Oil and Gas Exploration of East China Sea Basin.* Nanjing: Nanjing University Press, 1997.

[10] Zerong Liu, Xiaoling Wang. Re-discussion on Hebei-Shandong broom-shaped tectonic system. *Journal of East China Petroleum Institute,* 1981 10(2): 3–16.

[11] Yuzhu Kang. *Paleozoic Marine Oil and Gas Fields in Tarim Basin.* Wuhan: China University of Geosciences Press, 1992.

[12] Zhenliang Shi, Wenlin Huan. An overview of intraplate earthquakes. *Recent Developments in World Seismology,* 1984 10: 7–9, 33–34.

[13] Qidong Deng, Tingdong Wang, Jianguo Li, *et al. Structural Models for the Development and Occurrence of Haicheng Earthquake.* Beijing: Geological Publishing House, 1978.

[14] Ogata, Y., K. Shimazaki, Transition from aftershock to normal activity: The 1965 Rat Islands earthquake aftershock sequence. *Bulletin of the Seismological Society of America,* 1984 74(5): 1757–1765.

[15] Paul Tapponnier, Peter Molnar. Active faulting and tectonics in China. *Journal of Geophysical Research,* 1977 82(20): 2905–2930.

[16] Shaoxian Zang, Jieyuan Ning, Lizhong Xu. The distribution of earthquakes, behavior of the subduction zone and stress state in the Ryukyu Island arc. *Acta Seismologica Sinica*, 1989 11(2): 113–123.

[17] Shaoxian Zang, Zhongliang Wu, Jieyuan Ning, Sihua Zheng. The interaction of plates around China and its effect on the stress field in China, Part 2: The influence of Indian plate. *Chinese Journal of Geophysics*, 1992 35(4): 428–440.

[18] Yuzhu Kang. *Petroleum Geology Features and Petroleum Resources in Tarim Basin*. Beijing: Geological Publishing House, 1996.

[19] Yuzhu Kang. *Petroleum Geological Features and Resource Evaluation in Northwest China*. Urumqi: Xinjiang Science and Medical Publishing House, 1997.

[20] Zhonghuai Xu, Suyun Wang, Yanxiang Yu. Inversion of plate boundary forces based on observed stress directions using the finite element method. *ACTA Seismologica Sinica*, 1992 14(4): 446.

[21] Yuzhu Kang, Xiyuan Cai. *Formation Conditions and Distribution of Paleozoic Marine Oil and Gas Fields in China*. Urumqi: Xinjiang Science and Medical Press, 2002.

[22] Liang, T., Jones, B. Deciphering the impact of sea-level changes and tectonic movement on erosional sequence boundaries in carbonate successions: A case study from Tertiary strata on Grand Cayman and Cayman Brac, British West Indies. *Sedimentary Geology*, 2014 305: 17–34.

[23] Dunbar, G. B., Barrett, P. J., Goff, J. R., et al. Estimating vertical tectonic movement using sediment texture. *Holocene*, 1997 7(2): 213–221.

[24] Lister, C. Tectonic movement in the Chile Trench. *Science*, 1971 173(3998): 719–722.

[25] Srivastava, G. S., Kulshrestha, A. K., Agarwal, K. K. Morphometric evidences of neotectonic block movement in Yamuna Tear Zone of Outer Himalaya, India. *Ztschrift für Geomorphologie*, 2013 57(4): 471–484.

[26] Browman, David L. Archaeology: Tectonic movement and agrarian collapse in prehispanic Peru. *Nature*, 1983 302(5909): 568–569.

Chapter 4

Major Tectonic Movements All Over the World

Abstract

This chapter discusses the major tectonic movements all over the world, including between the Mesoproterozoic and the Cenozoic eras, between the Silurian and Devonian (the first act of the Hercynian Movement including the Tianshan Movement), between the Middle and the Late Triassic, between the Jurassic and the Cretaceous, and between the Paleogene and the Neogene. In the end, the chapter touches on the evolutionary profile of representative ultra-long geological structures all over the world.

Keywords: Global, tectonic movements, ultra-long geological structure, profile.

The structural evolution of the continental lithosphere was mainly caused by tectonic movements and the tectonic system of the continental crust. Their formation, evolution, and distribution law of the lithosphere were restricted by the crustal movement, different tectonic systems, and the directions of the movements. Therefore, the nature, era, and changing law experienced due to the main crustal movements shall be discussed from the perspectives of formation and evolution of the continental crust structure.

Crustal movements are often divided into orogenic movements and oscillatory movements. The scope they cover is different [1]. Tectonic movement even in the same period showed different performances in different geological contexts. For example, a movement created mountains in some areas due to squeezing or strike-slip uplifting, while in other areas, the land sank continuously due to depression or extensional subsidence. Hence, during late deposition, movements resulted in unconformable contact in some cases and continuous deposition or no obvious discontinuity in other cases. This was because, in crustal movement, compression and tension, fold uplift, and depression or fault depression always complemented or accompanied one another. However, as structural deformations such as compression and tension caused by tectonic movement had different products, extrusion or compression-shear deformation zones often saw accompanying deformation, as well as metamorphism and magmatic intrusion activities. Extensional rift zones were often accompanied by volcanism, such as magma eruption, forming active ocean troughs and volcanic-clastic rock formations, while depression areas or oscillation zones often had relatively stable clastic-carbonate formations, which played an important role in controlling sedimentary minerals or volcanic-sedimentary minerals. Late tectonic activities sometimes changed the original structure pattern and showed obvious newness. They sometimes also inherited part of the earlier structure pattern and showed obvious inheritance, related to the boundary conditions of each affected part and the distribution and change of tectonic stress fields during tectonic movement (Table 1).

The tectonic evolution history of major continents was different, and the manifestations and strengths of tectonic movements in each period were quite different in each continent as well. The North China landmass consolidated after two tectonic movements at the end of the Neoarchean and the Paleoproterozoic, forming a crystalline substrate. The Middle–Neoproterozoic was the development stage of stable caprock. It was not until the Indosinian Movement in the late Triassic that deformation and metamorphism with high activity intensity appeared [2]. The Yangtze–Talimu landmass formed a crystalline basement at the end of the Paleoproterozoic. During the Mid-New Proterozoic, it was an active and sub-active sedimentary environment. After the Jinning movement, a fold basement was formed. Tectonic evolution entered the stage of stable caprock development from the Sinian era. Since the Phanerozoic, the development history in this tectonic continental landmass remained

Table 1.　A summary of Africa's stratigraphic age, geological movements, and global comparison.

Geological age			Isotope/age value/Ma Present	Major geological events	Tectonic stage and crustal movements		
					Europe and America	China	Africa
Cenozoic	Quaternary period	Holocene	0.01	Disintegration stage of the united continent	New Alpine stage { Saff Movement, Pyrenean Movement }	Himalayan stage { Himalayan Movement (Late), Himalayan Movement (Early) }	Alpine stage { Late Alpine Movement, Early Alpine Movement }
		Pleistocene	2				
	Neogene period	Pliocene epoch	5				
		Miocene	22.5				
	Paleogene	Oligocene	37.5				
		The eocene epoch	50				
		Paleocene	65				
Mesozoic	The cretaceous period		137		Alpine stage { Laramide Movement, The New Simerian Movement, The Ancient Simerian Movement }	Yanshan Movement { Yanshan Movement (Late), Yanshan Movement (Middle), Yanshan Movement (Early), Indosinian movement (early), Indosinian movement (early) }	Hercynian Stage { Third act of Hercynian Movement }
	Jurassic Period		185				
	Triassic period		230			Indosinian Hercynian stage	Second act of Hercynian Movement
Neopaleozoic	Permian		280	Formation stage of united ancient land	Hercynian stage { Appalachan Movement }	Yiming movement	First act of Hercynian Movement
	Carboniferous		350			Tinuslau Movement	
	Devonian		400		Breton Movement		
Early paleozoic era	Silurian period		440		Caledonian Movement { Erian Movement, Taikang Movement }	Caledonian Movement { Qilian (Guangxi) Movement, Gulang Orogeny, Xingkai Orogeny }	Pan–African phase { (Katanga) Late Pan-African Movement }
	Ordovician period		500				
	Cambrian period		610				
Proterozoic	Sinian period		850		Anita Movement	Jinming Movement (Late)	(Katanga) Early Pan-African Movement
	Neo- -over- -zoic		1055	Stage of platform formation	Goethe and Greenville Movement	Luliang lining stage { Jinming Movement (Early), Luliang (Middle) Movement }	
	Meso- -zoic		1600–1700		Cary Hudson Movement		
	Paleop- -zoic		2500–2600			Wutai Orogeny	
Archaean	New		3900–3000	Stage of continental core formation	Sam–Kennel Movement	Fuping Changli stage { Fuping Movement }	
	Ancient		3800				
Preatrchaean			4600	Astronomical phase			

basically the same, except that the activity intensity in the west was sometimes greater.

Since Indosinian Movement, there were obvious differences between the east and the west. The Cathaysia Paleo continent formed a crystalline basement at the end of the Paleoproterozoic, which was in an active depositional environment from the Mesoproterozoic to the Early Paleozoic. That is, there was active sedimentation before and after the Jinning Movement, forming a fold basement in the early and late Paleozoic, followed by stable cap deposits. There was strong activity since the Indosinian Period [3]. The southern Xizang-Western Yunnan area was part of the northern margin of the Gondwana continent and was similar to the Cathaysia landmass. Its fold basement was formed in the Early Paleozoic, which evolved into the main component of the East Tethys tectonic domain during the Mesozoic. It was very active. The northernmost part of China was part of the southern margin of the Mongolian landmass, where the outcrop of basement rock formation was negligible. Data available indicated that the area may have been a vast ocean during the Paleozoic Period, whose folds returned at the end of the Paleozoic era when it entered a stable development stage. A Proterozoic crystalline basement may have existed. However, during the Mesoproterozoic, the Jiamusi–Laoyeling, Jidong, and Liaodong regions were uplifted for a long time and strongly subsided in the Neoproterozoic, which was consistent with the Dabie–Jiaodong region. Since the Neoproterozoic, this was similar to the Yangtze–Cathaysian landmass, but obviously different from the North China Landmass.

The main tectonic movements from old to new in the Paleozoic are mentioned in Table 1.

1. Tectonic Movements between the Mesoproterozoic and the Cenozoic

The Jinning Movement originally refers to an important tectonic movement in which the Kunyang Group and Huili Group in Southwest China deformed to form fold basements. It was found in subsequent work that this movement was widespread in the Yangtze landmass and its periphery, and was also an important cause of deformation and metamorphism that formed the basement of this area. In a later period, the area had extensive magmatic activities and transitional-active constructions, and was the location of the subsequent Chengjiang Movement or the tail of the Jinning Movement; there was a tectonic change between the Sinian glacial rock formation and the underlying rock formation, between 1,000–800 Ma [4].

This tectonic movement was evident in the Tarim–Qaidam Landmass, Songpan Landmass in northwestern Sichuan, Qiangtang Landmass in northern Xizang, the portion from the southern Alashan Landmass to the southern North China Landmass and the Funiu Mountain uplift belt area, which was called the Altun Movement, and the Tarim Movement or Quanji Movement in northwest China. There were two time periods, with the main time period being about 1,000 Ma and the later time period being approximately 800 Ma, which were roughly equivalent to the beginning and ending extremes of the Qingbaikou Period. Sedimentary construction, tectonic deformation, metamorphism, and magmatic activity of the fold base of the Yangtze–Talimu landmass were comparable and the Qingbaikou volcanic-clastic and flysch-like structures in these areas were formed in the same tectonic environment. The Sinian glacial diagenesis, carbonate rock formations, and the late phosphorus (velamen and uranium) formations also occurred. In general, the Jinning–Talimu Movement consolidated the basement folds of the Yangtze–Talimu landmass, after which it entered the stage of stable caprock development. It also had an important impact on the North China landmass, which mainly manifested in the overall uplift of the Sinian with little to no deposition from the period of 800–600 Ma from the present [5]. The Yangtze–Talimu landmass and the North China landmass formed the early stable and unified continental crust of China. The deformation domain of the Jinning–Talimu Movement extended into the southwest of the North China landmass. Beishan Mountain, South Alxa Mountain, and the front part of Funiu Mountain significantly deformed and metamorphosed during the Pre-Sinian and the Mid-New Proterozoic and became the northern margin of the Yangtze–Talimu landmass, and then later became the Sinian sedimentary area. This belt may extend northward to the eastern margin of the Junggar Basin.

2. Tectonic Movements between the Silurian and the Devonian (The First Act of Hercynian Movement)

The Hercynian was an important tectonic movement at the end of the Early Paleozoic and was widespread on all continental crusts. However, the intensity, manifestation, time period, and duration of tectonic movement varied in different landmasses or zones of the same landmass. Thus, the names of the different landmasses and regions are not uniform [6]. The representative ones are the Qilian Movement in northwest China and Guangxi Movement in south China, which featured fold deformation in the late early Paleozoic. The North China landmass was characterized by uplift and denudation from the Late Ordovician to Early Carboniferous.

In the periphery of the Cathaysia landmass and the Yangtze landmass and some internal active zones, a fold movement occurred at the end of the Silurian causing strong deformation to the Lower Paleozoic rocks, accompanied by varying degrees of dynamic metamorphism, intrusion of intermediate-acid magma, significant fault activity, and landmass uplift. For example, in the North Qinling Belt, the Longmenshan–Yulong Snow Mountain Belt, and the eastern margin of the lower Yangtze and Cathaysian landmass, tight linear folds and large fault zones were formed. In the Ailao Mountain Belt and the Ziyun–Luodian Fault Fold Belt in the western and southwestern margins of the Yangtze landmass, there was significant unconformable contact between the Devonian and the Lower Paleozoic, which resulted in the Sichuan Basin rise and general rise of the neighboring Dabashan and Daloushan. But, there was no obvious deformation. It featured erosion unconformity during most of the Middle-Lower Devonian and Silurian. It was the same case within the Lower Yangtze landmass.

3. Tianshan Movement (The First Act of the Hercynian Movement)

The Tianshan was an important tectonic movement in the middle and late Paleozoic. Due to the intense transformation of the Chinese continental crust by the Caledonian Movement, the Cathaysia landmass and the Yangtze landmass joined together, which changed the boundary conditions of eastern and southeastern China. In addition, as the Indian landmass was close to the Tarim–Yangtze landmass, the formation and development of the Mongolian–Caledonian Arcuate Tectonic Belt gradually changed the pattern of the crust structure of Chinese continent since the Late Paleozoic. In the early part of the Late Paleozoic, China's continental crust generally underwent a transition from uplift to stable sedimentation, which had inheritance activities with frequent ups and downs in some Caledonian active zones. The Devonian system saw most piedmont-intermountain continental or continental–coastal shelf clastic rock formations and basically no normal marine carbonate formation, which was missing in many areas. The Late Hercynian Movement mainly occurred between the Early and Late Permian. The Early Permian featured fold deformation and magmatic activity, with significant unconformable relations, whereas the Later Permian was characterized by intermittent uplift, with partial unconformable relations [7]. The North China landmass featured only weak discontinuities. The Tianshan Movement showed that the two zonal belt systems that appeared since the middle and late periods of the Late Paleozoic gradually formed from west to east and deformed to form a fold belt with strong magmatic intrusion, which was covered by the late Permian stable clastic-carbonate rock tectonic unconformity. During this period, there were also several plateau basalt belts in the north–south direction. The most notable were the Emeishan basalt belt along the north–south direction of the Sichuan–Yunnan area, the basic intrusive rock belt containing vanadium, titanium, and magnetite in the Panzhihua–Xichang area, and the Jinshajiang–Lincang tectonic magmatic dynamic metamorphic belt.

It should be pointed out here that the tectonic movements of the continental crust of China in most places ended not between the late Permian and Triassic but in the early and late Permian, while the late Permian and early Triassic saw mostly continuous deposits, without any unconformity.

4. Tectonic Movement during the Middle and Late Triassic

There was strong tectonic movement occurring in the Tertiary Period, and its main act occurred in the Middle First Late Triassic or the Late Middle First Triassic, which moved the evolution of the tectonic system between the Asian continent and the Pacific landmass to a new stage [8]. There were strong and obvious traces in the continental crust of China, or the East Asian region bordering the Pacific Ocean and the Indosinian region.

The most prominent changes were the development and finalization of two zonal tectonic systems that traversed central China, the rise of the Tethys tectonic belt in southwestern China, the formation of the Eastern Cathaysia structure, and the formation of the Indosinian active tectonic belt on the southern margin of the continent. As a result, the deposits from the Late Triassic Retician to the Early Jurassic Riasian were extensively unconformable on the strongly deformed metamorphic rock formations of the middle and late Triassic. Except for the western Tethys Basin which evolved into an active marine sedimentary environment, in the rest of the Chinese continent, the transgression ended and the environment turned into continental coal-bearing sediments after orogeny. After the redeformation of the Indosinian tail at the end of Late Triassic, the southwest seawater continued to retreat westward to a corner of southern Xizang. Acid magma intrusions and tectonic dynamic metamorphism widely developed in the Indosinian Movement.

After the strong and extensive Indosinian Movement that the Chinese continental crust experienced, the present unified continental crust was basically formed and the boundary conditions between the three major structural domains became a certainty. The subsequent Yanshan Movement and Himalaya Movement mostly transformed and developed on the basis of the Indochina framework, with only some local or regional changes. Therefore, the Indosinian Movement was another major change in the evolution of the Chinese continental crust and was comparable to the Luliang Movement and Jinning Movement [9].

5. Tectonic Movements between the Jurassic and Cretaceous

It is generally believed that the Yanshan Movement was an important tectonic movement that widely developed throughout China from the Jurassic to the Cretaceous, which mainly manifested as fold and fault movements, magmatic intrusion and eruption, and metamorphism in some areas. It was not only an important tectonic movement of the Chinese continental crust but also had a great influence on the Pacific Rim Region and the Middle Tethys tectonic domain [10]. As the features and deformation intensity of this movement in various regions were different, there were some differences in the division of its tectonic periods, which were generally divided into three strong fold-fault deformation periods and two weaker deformation periods (a total of four acts: Middle-Late Jurassic, Late Jurassic–Early Cretaceous, Early-late Cretaceous, and Late Cretaceous–Paleocene), with the magmatic activity and structural deformation between the Late Jurassic and Early Cretaceous being the most evident. There were however still major differences in the movement features and expression modes between the Late Cretaceous and the Paleogene. As the sedimentary environment in the Late Cretaceous and the Paleogene was basically the same and the stratum was mostly continuous or without obvious discontinuities, there was no regional unconformity [11].

The Yanshan Movement in China was actually the continuation and development of the Indosinian Movement, which changed from the north–south equilibrium squeeze of the Valixi–Indosinian Period to an unbalanced environment of squeezing and twisting and was dominated by non-equilibrium twisting in the Yanshan Period. The deformation characteristics changed from strong plasticity to brittle plastic deformation. The Yanshan Movement not only produced new structural patterns but also strengthened and inherited some early structural types and patterns and cast the structural features of the current China continental crust.

6. Tectonic Movements between the Paleogene and Neogene

There was tectonic movement that occurred in the Chinese continental crust since the Cenozoic, which turned the Mesozoic Tethys sea area into a huge mountain range and resulted in the formation and development of the ditch-arc basin near the Pacific Ocean. Its main deformation events occurred in the beginning of the Oligocene and the Miocene. Since the Oligocene, the seawater in Xizang completely withdrew and violent deformation and metamorphism followed, which featured strong folds, faults, neutral–acid magma intrusion, and dynamic thermal metamorphism, with large-scale thrusting, overthrusting, and sliding in the later period. As a result, the crust of Qinghai–Xizang Plateau was greatly uplifted and extended from east to west, which led to lateral migration of materials, causing the eastern continental crust to rotate clockwise around the Qinghai–Xizang landmass, making the Pamir–Himalayan region the highest and youngest wrinkled mountain system in the world [12].

Constrained by the Pacific tectonic domain, the Taiwan province of China–Philippines fold mountain system along the north–south direction was formed in eastern China in the Neogene and early Quaternary, and the NE–NNE structural belt and East Asian island arc belt on the southern continental shelf of China further developed.

The first act of the Himalayan Movement took place between the Neogene and the Paleogene, leading to strong unconformable contact between the Neogene and the Paleogene. The second act happened between the Miocene and the Pleistocene. The southeastern waters and the Taiwan province of China–Philippines area, part of the Taiwan Movement, reached a stage of violent activity before the Pleistocene. The third act lasted from the Pleistocene to the present. The western plateau uplifted sharply under the action of north–south squeezing and unbalanced twisting stress. Old faults became active again. Quaternary volcanic eruptions occurred in some areas and strike-slip activities were obvious in the eastern margin. Under the influence of the Pacific landmass, the eastern region suffered from east–west compression, which produced a right strike-slip along the early NE–NNE trending fault zone. Meanwhile, a series of secondary pull-apart basins were also formed, which changed the mechanical nature and mode of movement of the early fractures [13]. This movement is still active even today and has an important control effect on seismic activity and other geological disasters in the region.

7. Evolution Profile of Global Representative Ultra-long Geological Structure

7.1. *India–Siberia–North America–South America Meridional Super-long Profile (A1–A6)*

The India–Siberia–North America–South America meridional ultra-long profile (Figure 1) started from the passive continental margin of the northwestern Indian continent and passed through the main Indian continent, the Tethys Orogenic Belt, Central Asia, Siberia, the Arctic Ocean, and the continents of North American and South American, with a total profile length of approximately 27,500 km. In the following, we discuss the profile from three aspects, namely, the Indian–Siberian continent meridian long profile (Profile A1–A2), the North American–Caribbean

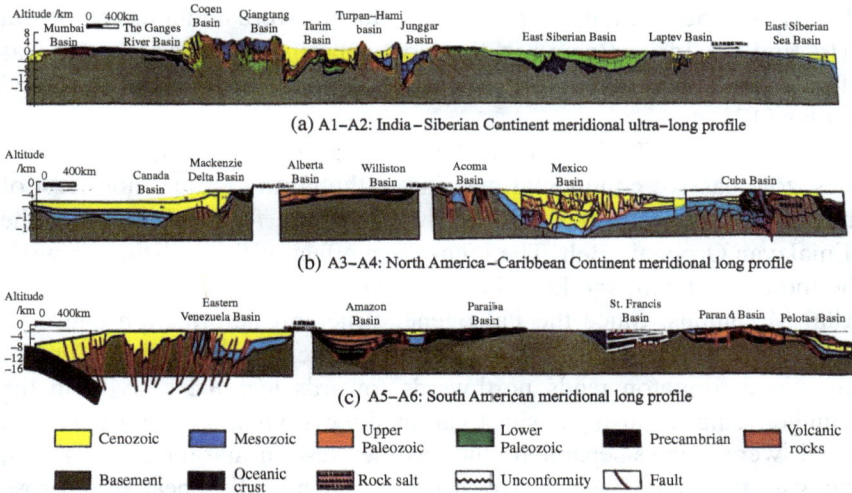

(a) A1–A2: India–Siberian Continent meridional ultra–long profile

(b) A3–A4: North America–Caribbean Continent meridional long profile

(c) A5–A6: South American meridional long profile

Figure 1.   India–Siberia–North America–South America meridional super-long profile (A1–A6).

continent meridional long profile (profile A3–A4), and the South America meridional long profile (A5–A6 profile) [14].

(1) **A1–A2: Indian–Siberian continent meridional long profile:** The Indian–Siberian continent meridional long profile has a total length of approximately 8,000 km, with the thickness of the caprock of different sedimentary basin groups being between 6 and 14 km [15]. From south to north, the profile passed through the Indian continent, the Lhasa landmass, the Qiangtang landmass, the Tarim landmass, the Siberian continent, and the Tethys and Central Asian tectonic domains. On the north and south sides were the Indian Ocean Basin and the Arctic Ocean Basin, respectively.

The profile extended from south to north and included, in order, the Mumbai Basin, the Deccan Basalt Province, the Ganges Basin, the Himalayan Orogenic Belt, the Qinghai–Xizang Plateau (including the Cuoqin Basin and the Qiangtang Basin), the West Kunlun Orogenic Belt, the Tarim Basin, Tianshan Basin, the Craton Basin (Siberian Basin), and superimposed basins (the Talimu Basin and the Junggar

Basin). Among them, the superimposed basins were under development [16], and the Tarim Basin had the most superimposed periods and was affected by the two major structural domains of Central Asia and Tethys [17].

The Cenozoic Indian continent collided strongly with Eurasia, and the Mesozoic passive continental margins in the northwest and northeast of the Indian Continent evolved into the peripheral foreland basins of the Himalayan Orogenic Belt. The long-range effect of the collision between the Indian continent and Eurasia can extend northward to the north of the Altai Mountains. Since the Paleogene, structural deformation and continuous northward spread of stress and tectonic activity, and the time of tectonic deformaton tends northwards towards tectonic activity in the Younger Ranges causing basins to gradually evolve into structural basins, which were always separated (such as the western margin of the Tarim, Junggar, and Tuha Basins). The basins constantly disappeared and rose (such as the Chaidamu Basin) and the basin area was constantly shrinking (such as Fergana Valley and the Afghanistan–Tajikistan Basin).

The Mumbai Basin was formed in the Early Cretaceous. After Madagascar was separated from the Indian continent, the passive continental margin deposition stage began in the late Paleocene and early Eocene, which led to the formation of a typical passive continental margin basin and one of the richest oil basins in South Asia. The Mesozoic Cuoqin Basin and Qiangtang Basin developed in the Tethyan tectonic domain and were greatly uplifted since the Paleogene, with the sedimentary strata structurally destroyed, forming the Qinghai–Xizang Plateau. Influenced by the revival, uplift, and structural expansion of Pamir and Tianshan Orogenic Belts, the Cenozoic tectonic deformation of the Tarim Basin spread from the edge to the basin to the surrounding orogenic belts, becoming strongly deformed and uplifted, which caused the Tarim Basin to suffer from Cenozoic deflection and subsidence. The Tarim Basin was shortened greatly in terms of structure and became an inland tectonic basin. The Mesozoic and Cenozoic basin profiles in the Turpan–Hami Basin were asymmetrical in shape and characterized by foreland basins; that is, they were close to the mountain front and had increased stratum thickness and increased structural deformation. The southern margin of the Junggar Basin was affected by the thrusting nappe of the Tianshan Mountains in the Cenozoic, forming a foreland basin. The Siberian Platform was mainly a Paleozoic sedimentary basin, which experienced two stages of aulacogen-passive continental margin formation during the

Liffey–Wende Period and the craton platform basin formation during the Paleozoic. It uplifted and eroded [18] after the Triassic and controlled the current landform development [19]. As it was far away from the boundary of the compressed plates, the tectonic activity of the Siberian Basin remained stable for a long time and the degree of exploration was very low.

(2) **A3–A6: North American–South American meridional profile:** The North American–South American meridian profile was approximately 19,500 km long in total. The main tectonic units it passed from north to south included the Canadian Basin, the Alaska Great Basin, the North American continent, the Appalachian–Woxito Orogenic Belt, the Caribbean continent, the Caribbean Peripheral Orogenic Belt, the South American continent, and the Atlantic passive continental margin. The profile also passed through the Canadian passive continental margin basin, the Mackenzie Delta Basin, the Alberta Foreland Basin, the Akoma Craton Basin, the passive continental margin basin of the Gulf of Mexico, the Cuban Fore-arc Basin, the Eastern Venezuelan Foreland Basin, the Amazon Craton Basin, the Parana Craton Basin, and the Atlantic Passive Margin Basin (Perotas Basin).

The maximum buried depth of Mackenzie Basin was 16 km. The basin was formed when the Canadian Ocean Basin opened, which was dominated by passive continental marginal deposits during the Paleozoic–Mesozoic era. The basin rose in the mid-Mesozoic era and the sedimentary gaps suffered weathering and denudation. It was dominated by delta deposits in the Cenozoic era. The passive continental margin of Alaska gradually stretched and deformed during the middle and late Eocene. In the Late Miocene, due to the surrounding uplift and denudation, deformation occurred in the southern part of the Mackenzie Basin [20].

The Alberta Basin was rich in oil and gas and was a retro arc foreland basin [21]. The basin was adjacent to the Mackenzie Basin in the north and the Williston Basin in the south. The Alberta Basin was gently inclined to the west in profile, with its wedge-shaped sediments gradually thinning from west to east. The Alberta Basin was the thickest in the west, up to 600 m. It extended eastward to the Precambrian basement. The basin was a passive continental margin basin in the Paleozoic, and a typical foreland basin in the Mesozoic and Cenozoic due to the Cordillera orogeny in western North America [22]. The Larami Tectonic Movement during the Mesozoic further expanded the foreland basin and deposited thick layers of clastic deposits on the continental margin due to which the deposition center of the foreland moved eastward.

The Akoma Basin was a Paleozoic craton basin. The Ouachita Movement in the late Pennsylvanian Period formed the Ouachita Forel and Thrust Belt in the southern part of the basin. The Gulf of Mexico Basin was a passive continental marginal basin, which extended since the Mesozoic and saw rapid subsidence and thick delta deposition. The deposition and subsidence center of the Gulf of Mexico Basin continued to move south from the Mesozoic to the Cenozoic, which was an intercontinental oceanic rift basin in the late Mesozoic after which the ocean basin collapsed. It is now a continental marginal basin.

The northern Cuban Basin was a fore-arc basin developed on the north side of the Arc of Grand Andres Island in the Caribbean Sea. The Arc of the Grand Andres Island was close to the Caribbean Sea, and developed a series of retro arc basins, such as the Central Cuba Basin, the Southern Cuba Basin, the Caribbean Coastal Plain Basin, and the Grenada Basin. The Grand Andres Island arc was affected by strike-slip and the abovementioned back-arc basin was in a shear-tensioned structural environment.

The Eastern Venezuela Basin was a part of oil- and gas-enriched basin in South America. It was a back-arc foreland basin that evolved from a passive continental margin basin, which had experienced pre-rift during the Paleozoic, initial rift during the Jurassic–Cretaceous Period, passive continental margin occurrence during the Cretaceous–Paleogene, and foreland basin evolution during the Neogene [23]. The East Venezuela Basin had an asymmetric structure with a wedge-shaped profile. The strata gradually thinned from the northern orogenic belt to the southern platform and sloped to the north. During the end of the Cretaceous and the Paleocene, the oblique collision of the Caribbean landmass and the South American continent caused the mountain system to uplift and the Flysch Trough Basin to migrate southward. When it came to the Eocene, the edge of the plate had become a right-handed strike-slip type, and continuous compression caused the loaded crust to thrust, which led to the evolution of the foreland basin in the Eastern Venezuela Basin, but the structural deformation was not strong.

The Amazon Basin was a Paleozoic craton basin which was mainly filled with and developed three sets of transgressive and regressive sedimentary structural sequences. The maximum sedimentary thickness of the basin exceeded 7,000 m. The Amazon Basin developed one basic structural framework each in the south and north. Between the two platforms was a central depression, representing the rift axis from the Precambrian to the Paleozoic. The oil and gas reserves were not high [24].

During the Early Cretaceous, a series of fault depressions developed in the eastern part of the South American continent, which developed into a passive continental margin basin from the Late Cretaceous to Neogene. The Pelotas Basin in southern Brazil experienced three evolutionary stages: pre-rift, fault depression, and post-fault migration. Like the rift evolution stage, block uplifts and volcanic activity occurred extensively, forming fluvial sand, conglomerate rocks, and volcanic rocks. At the beginning of the migration and evolution stage of the fault depression, argillaceous limestone and sandy limestone were abundant. Subsequently, thick clastic rocks were deposited during the lower Cretaceous Series forming the Paleogene shelf and shelf slope.

7.2. *North America–North Africa–Middle East–Central Asia–East Asia Zonal Ultra-long Profile (B1–B6)*

The North America–North Africa–Middle East–Central Asia–East Asia zonal ultra-long profile (Figure 2) started from the east coast of the Pacific Ocean and passed through the North American continent, the northern part of the African continent, the Middle East, Central Asia, East Asia, and the Western Pacific in turn from west to east. It was approximately 19,000 km in length and divided into three parts: the North American zonal long profile (B1–B2 section), the North African zonal long profile (B3–B4 section), and the Middle East–Central Asia–East Asia zonal long profile (B5–B6 profile).

(1) **B1–B2: North American zonal long profile:** The profile was approximately 2,000 km long in total and the caprock thickness of different sedimentary basin groups ranged from 3~12 km. It passed through the Los Angeles Basin, the Rocky Mountain Basin Group, the Williston Basin, the Illinois Basin, the Appalachian Basin, and the East Coast Basin from west to east. Affected by the subduction of the East Pacific continent to the North American continent on the east side of the profile, back-arc foreland basins developed during the Mesozoic and Cenozoic. The central part of the profile contained stable cratonic deposits

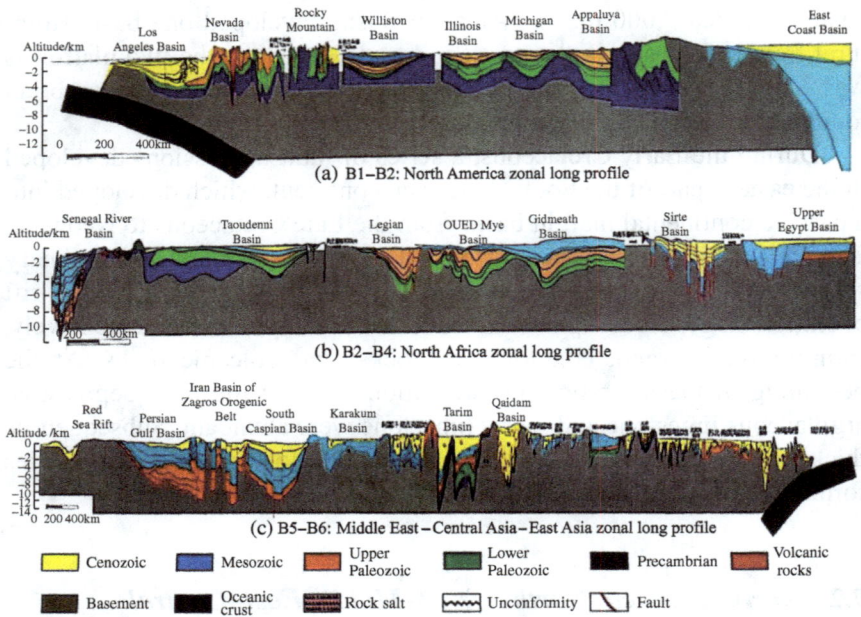

(a) B1–B2: North America zonal long profile

(b) B2–B4: North Africa zonal long profile

(c) B5–B6: Middle East – Central Asia – East Asia zonal long profile

Figure 2.　North America–North Africa–Middle East–Central Asia–East Asia zonal ultra-long profile (B1–B6).

since the Phanerozoic. The Appalachian Basin in the west section of the profile was the Late Caledonian foreland basin of the Early Paleozoic and developed a series of thrust faults thrusting westward [25]. Passive continental margin basins developed in the Mesozoic and Cenozoic era because the North Atlantic Ocean was opened.

The Los Angeles Basin was a pull-apart basin formed under the control of the San Andres Great Fault and is the most abundant basin of oil and gas per unit area in the world. The basement of the basin was composed of metamorphic rocks and intrusive rocks during the Late Jurassic and Early Cretaceous, on which strata were deposited from the Upper Cretaceous to Pleistocene; the maximum thickness of the strata in the center of the basin reached 9,400 m. The Nevada Mesozoic Rift Basin developed on the Cordillera Orogenic Belt and the Colorado Plateau in western United States and was composed of multiple north–south fault depressions, forming a basin and range province. The sedimentary cover of the Rocky Mountain Foreland Basin Group was formed from the Precambrian to the Neogene, bounded by the Cordillera Overthrust Nappe-fold Belt to the west and connected to the Alberta Basin of Western

Canada in the north. The formation and evolution of the basin were related to the subduction and compression of the East Pacific Plate toward the North American Plate (approximately 140 Ma).

The Illinois Basin was elliptical and was a gradual development of the Craton Basin on a complex of wanting faults and depressions (Early Carboniferous in the Cambrian) [26]. The thickness during the Upper Cambrian Series and Pennsylvania System reached 4.3 m. In the Early Cambrian, it was a rifted basin and turned into a craton sag in the Late Cambrian–Permian. In the middle Paleozoic, it connected with the passive continental margin in the east and marine carbonate rocks were deposited. Affected by the eastern Oujito and Appalachian tectonic movements, it suffered compression and transformation in the late Paleozoic. Most of the large anticlines and faults in the basin were formed or activated during this period. During the Mesozoic and Cenozoic Eras, the structural subsidence of the Illinois Basin ceased and it was uplifted and eroded. It was detached from the Appalachian Basin by uplift.

The Michigan Basin was in the shape of a disc, with strata deposited from the Cambrian to the Permian. It was dominated by carbonate deposits, with a maximum thickness of approximately 500 m. The Appalachian Basin extended northwestward, was located on the west side of the Appalachian Orogenic Belt formed by the Tacan Movement, and was detached from the Michigan Basin and the Illinois Basin by the continental emergence. The Paleozoic stratum developed, which sloped eastward and was thicker in the east and thinner in the west, with local thickness exceeding 12,000 m. In terms of evolution, the basins experienced the Cambrian rift stage, the passive continental margin stage during the Early Ordovician–Middle Ordovician, and the foreland basin stage during the Late Middle Ordovician–Pennsylvanian.

(2) **B3–B4: North Africa zonal long profile:** The profile was approximately 5,000 km long in total and the caprock in different sedimentary basin groups was 3–14 km thick. From west to east, it passed through Senegal's passive continental margin basin, the Taoudenni Craton Basin, the Legian Craton Basin, the Oued Mye Basin, the Sirte Rift Basin, the Upper Egypt Rift Basin, and the Red Sea Fault Basin. Different cratonic basins were separated by continental arch structure that developed in the Mesozoic. The Paleozoic craton basin in North Africa experienced two evolutionary stages: the Sinian Rift System and the Paleozoic craton basin [27]. Superimposed basins developed in North Africa in the Paleozoic. Since the Mesozoic, African continental sedimentary basins

have been dominated by uplift and denudation, providing material sources for peripheral passive marginal basins.

The Paleozoic craton basin in North Africa has similar geological characteristics. The super basin appeared in the early Paleozoic, which was composed of a strongly extended craton basin and a passive continental margin basin on the north side. It was related to the expansion of the Paleo-Tethys Ocean and influenced by the Late Paleozoic Wahito and the Allegheny Orogenic Movement (Hercynian Orogenic Movement). It experienced structural destruction, uplift, and erosion. The Sirte Basin was located at the northeastern end of the West African fault-depression system (Early Cretaceous) and was a Mesozoic and Cenozoic faulted basin. It was an important oil-bearing basin in Africa with a fault-depression dual structure [28].

(3) **B5–B6: the zonal long profile of the Middle East–Central Asia– East Asia:** The profile was approximately 12,000 km long in total and passed through the Red Sea Rift, Arabian continent, Zagros Orogenic Belt, Iranian landmass, Karakum continent, Pamir Orogenic Belt, Tarim continent, Qaidam continent, Qilian Orogenic Belt, North China continent, Northern Jiangsu Orogenic Belt, Yangtze continent, Ryukyu Trench, and other tectonic units from west to east. The basins it passed through from west to east mainly included the Red Sea Rift Basin, the Persian Gulf Superimposed Basin, the Southern Caspian Sea Remnant Ocean Basin, the Karakum Basin, the Tarim Basin, the Qaidam Basin, the Corridor Basin Group (including the Jiuxi Basin, the Huahai Basin, the Jiudong Basin, the Lake Basin, and the Bayanhot Basin), the Ordos Basin, the Taiyuan Basin, the Qinshui Basin, the Bohai Bay Basin, the North Yellow Sea Basin, the South Yellow Sea Basin, the East China Sea Basin, and the Okinawa Trough. It covered multiple basin types such as superimposed basins, craton basins, rift basins, foreland basins, and back-arc basins. Among them, the Persian Gulf Basin, the South Caspian Basin, and the Primorye Caspian Basin together constituted the world's most oil- and gas-rich area.

The Red Sea Rift Basin opened in the Neogene and was related to the activity of the Afar Mantle Plume. Since the Eocene, the impact of the Arabian plate collision on the interior of the Eurasia has been weak, mainly resulting in the structural inversion of the Central Iran Basin. The prototype of the basin on the profile was preserved intact, the fault activity was weak, and the basin continued to sink and deposit. The Persian

Gulf Basin was a passive continental margin basin superimposed on the Early Paleozoic–Devonian craton basin during the Permian–Jurassic Period, and it was a compound basin superimposed in the north by the Zagros Neogene–Quaternary foreland basin [29]. The Persian Gulf Basin had been in the passive marginal environment of the Paleo-Tethys Ocean and Neo-Tethys Ocean in Gondwanan continent for a long time and was marginally transformed by the Hercynian Orogeny and Himalayan Movement. The Zagros Piedmont fold and thrust belt, Arabian platform slope belt, Mesozoic evaporite basin, and Persian Gulf were particularly rich in oil and gas resources, forming a series of first-rate giant oil and gas fields. The Central Iran Basin, developed on the basement of an island arc complex, was a Mesozoic rift basin with Cenozoic structural inversion. The buried depth of the basement of the southern Caspian Sea was 20–25 km and it was one of the deepest sinking basins in the world.

The Karakum Basin (the Amu Darya Basin) and the Afghanistan–Tajik Basin developed on the Hercynian basement. The Amu Darya Basin experienced an early fault-depression (Late Triassic–Paleogene) and the fore-land basin stage (Neogene–Quaternary). During the Late Jurassic, it developed a salt structure. The basin's profile was asymmetrical, with deep faults, a stable basin structure in the middle stage, and thrust deformation in the southern margin during the late stage. Slight fold uplift occurred with weak structural transformation and small basin subsidence during the Cenozoic [30]. The Afghan–Tajik Basin was affected by the long-range compression of the Indian plate since the Paleogene. Because of mountain revival and tectonic uplift, it was compressed and shortened and became a foreland basin. Affected by the uplift and structural expansion of the Pamir Orogenic Belt, it saw relatively large Cenozoic deflection and experienced strong structural shortening. The basin was destroyed, uplifted, and eroded, and hills and highlands developed.

The Tarim Basin developed on the Tarim continent and deposited multiple sets of strata. Its profile shape showed characteristics of three folds and two uplifts. The western part of the basin detached from the Karakum landmass due to the Pamir Terrane and the Kunlun Mountain Orogenic Belt. The Middle-Upper Paleozoic strata in the Kunlun Mountain Orogenic Belt indicated that the Tarim landmass and its neighboring landmasses completed aggregation at the end of the Late Paleozoic. Later, due to the closure of the Neo-Tethys Ocean, foreland inversion occurred in the Tarim Basin, which also caused the reactivation and uplift of surrounding orogenic belts. At present, the Tarim Basin is characterized as an "enclosed basin" [31]. The Qaidam Basin and the Corridor Basin Group were intermountain basins sandwiched between the Paleo-Asian Ocean tectonic domain and the Paleo-Tethys tectonic domain. The Jurassic–Quaternary system was deposited in the basin. During the formation of the pan-continent in the Triassic, the basement of the Qaidam Basin and the Corridor Basin Group was formed. The basin groups in the middle section of the Jurassic all developed with extensional faults and foreland reversal because of the Himalayan Movement. Nowadays, thrust faults have developed in the basin and the Jiuquan Basin shows obvious characteristics of early fault depression and late foreland transformation. The basin groups that developed on the North China continent, the Yangtze continent, and the East China Sea continent all have the basement of the Early Precambrian craton. Since the Triassic, the Ordos Basin has been characterized by subduction in the foreland of the Helan Mountains. The internal tectonic activity in the basin was relatively weak and faults developed in the western margin. The Qinshui Basin was characterized by a huge Paleozoic syncline, with Neogene strata developed inside. The Bohai Bay Basin and the East Basin Group were strongly influenced by the Pacific Ocean subduction to Eurasia, which manifested as a dual structure of fault and depression, with the fault depression occurring gradually toward the east earlier. The Okinawa Trough was characterized by a single fault depression, and its formation was related to the subduction and back-arc extension of the western Pacific Ocean, forming a transitional crust.

Comparing the tectonic evolution of the global zonal ultra-long profile, it can be seen that basins on continental landmass and stable continents had a long evolution history and a stable tectonic history, while basins on the orogenic belts and micro-land belts (Central Asia) had a short evolutionary history and were easily interrupted. Except for the Paneba and St. Francis Basins on the South American continent that have developed faults, tectonic movement before the Precambrian mainly formed the basement of basins on each stable continent. In the Early Paleozoic, basins on stable continents were characterized by fault depressions (North American and African basin groups) and intracratonic depressions (the Tarim and North China Basin groups). The Caledonian Movement at the end of the Early Paleozoic uplifted and eroded the Williston Basin on the North American continent. The Appalachian Basin on the continental margin of North America, due to the orogeny at that time, evolved into the foreland basin development stage from the passive continental margin basin stage. Basins on the Central Iran landmass and the North China landmass also showed uplift and denudation characteristics. The Hercynian Movement in the early period of the Late Paleozoic manifested as the development stage of the uplift and denudation of the Asian Siberian landmass, the Tarim landmass, and the North China Continental Basin. In the Paleozoic, the Paleozoic basement of the Central Asian basin group was formed during the closure of the Paleo-Tethys Ocean. The Indosinian Period was the main period for the collision of many microcontinents in the northeast of Mainland China, North China, South China, and Mongolia. The Tarim, Ordos, and Qinshui Basins on the North China landmass and other basins showed characteristics of foreland basins or uplift and denudation [32]. During the Indosinian Movement, the Okhotsk Ocean finally closed and the Chinese–Mongolian continent accretion collaged in the southern margin of the Siberian continent. With the final closure of the Neo-Tethys Ocean and the development of the Alpine–Himalayan giant collision orogen, the Central Asian Basin Group and the Middle East Basin developed the characteristics of foreland basins. During the formation of the Eurasia, subduction always occurred around the Pacific Ocean. The subduction of the West Pacific Block caused the basins on the North China landmass to develop the characteristics of back-arc faults and depressions. The East Pacific continent subducted toward the American continent, causing the Rocky Mountain Basin Group in western North America and the Chaco–Parana Basin in South America to develop foreland characteristics. At the same time, during the disintegration of the pan-continent, the opening of the Atlantic Ocean led basins in eastern North America and western Africa to develop passive continental margins.

7.3. *Extra-long Sectional Profile of the East Coast of Africa–Mediterranean Sea–Europe–Arctic Ocean Coast–Siberia–Australia (C1–C8)*

The ultra-long sectional profile of the east coast of Africa–Mediterranean–Europe–Arctic Ocean coast–Siberia–Australia (Figure 3) ran from south to north through the east coast of Africa–Mediterranean–Europe and passed south through Siberia–China–Southeast Asia–Australia after running through the Arctic Ocean, with a total length of about 20,000 km, including the meridional long profile of the east coast of

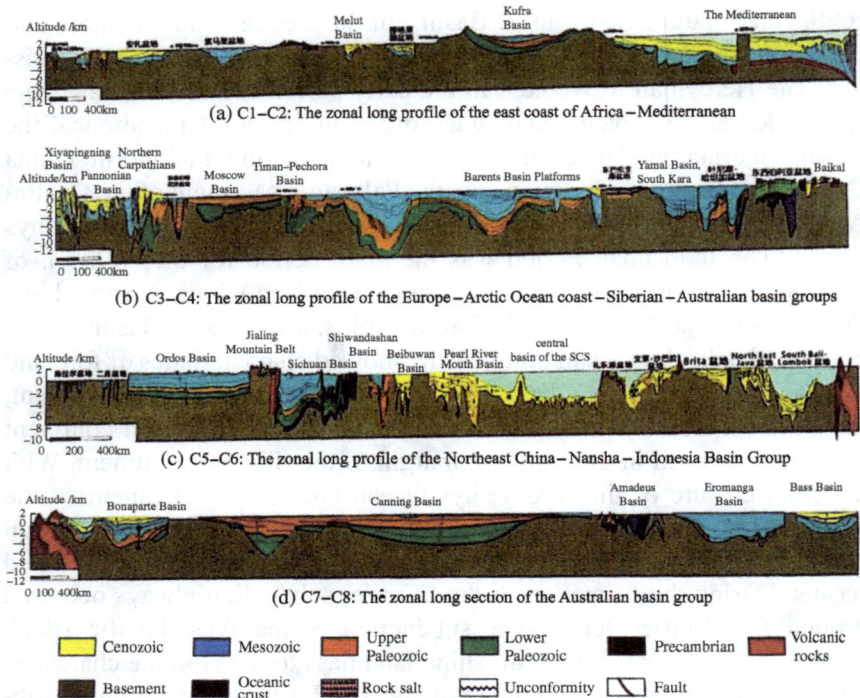

(a) C1–C2: The zonal long profile of the east coast of Africa – Mediterranean

(b) C3–C4: The zonal long profile of the Europe –Arctic Ocean coast –Siberian –Australian basin groups

(c) C5–C6: The zonal long profile of the Northeast China – Nansha – Indonesia Basin Group

(d) C7–C8: The zonal long section of the Australian basin group

Figure 3.   The ultra-long sectional profile of the East Coast of Africa–Mediterranean–Europe–Arctic Coast–Siberia–Australia (C1–C8).

Africa–Mediterranean basin group–Europe–Arctic Ocean coast–Siberian basin group and the meridional section of Northeast China–Nansha–Indonesian basin group–Australia.

(1) **C1–C4: Meridional long profile running through the east coast of Africa–Mediterranean Basin Group–Europe–Arctic Ocean Coast–Siberian Basin Group:** The Africa–Europe meridional profile passed from south to north through the African continent, the Alpine Orogenic Belt, the Valissian Orogenic Belt, the Eastern European Plate, the Ural Orogenic Belt, the Kara continent, and the Temel Orogenic Belt, involving the passive continental margin basins of East Africa, the African Meerut Strike-Slip Basin, the Muglad Rift Basin, the Khufra Craton Basin, the Eastern Mediterranean Remnant Ocean Basin, the Pannonian Intermountain Basin, the Carpathian Foreland Basin, the Dnepr–Nenets Craton Basin, the

Moscow Craton Basin, the Timan–Pechora Foreland Basin, the East Barents Craton Basin, and the South Kara Sea Craton Basin.

Sedimentary basins in eastern Africa were mainly controlled by the Central African fault system and had a series of Mesozoic–Cenozoic strike-slip-fault basins, including the Dopa Basin fault depression, Doceo fault depression, Salamat fault depression, Sudan fault depression, and Anyang fault depression, Anza fault depression, and Ramu fault depression. The Muglad Basin was the largest oil-bearing Mesozoic faulted basin in the faulted basin group in Sudan and was located in the south of the eastern end of the Central African fault system. Its plane shape was wide in the north and narrow in the south. In addition, it was distributed in a long wedge shape and developed with Cretaceous–Paleogene sediments. A shallow dish-shaped craton basin developed in northern Africa. The passive continental margin and oceanic crust in northern Africa subducted northward in the eastern Mediterranean, forming a residual oceanic basin in the eastern Mediterranean.

The final consolidation of the western continental crust basement was related to the basement development of the Caledonian and Hercynian Orogenic Belts. Eastern Europe developed the Mesozoic craton basins in the northeast, comprising a huge area and sedimentary thickness and weak structural deformation. The Alps–Carpathian Orogenic Belt was the product of the collision between the African continent and Eurasia during the Late Cretaceous–Paleogene [33]. The Carpathian Fold Belt was the result of the uplift of the Pannuo landmass toward the Eurasian Plate during the Alpine Period. The forepart of the Carpathian foreland thrust belt formed the southern and northern Carpathian basins; that is, the Panno Basin and the Ransivania Basin were formed in the intermountain landmasses, both of which were rift sedimentary basins controlled by mantle uplift after the basic formation of the Alps fold belt. The Dinara and Apennine Fold Belts were on the south side of the Alps, spreading from northwest to southeast. The Dinara Fold Belt thrust from northeast to southwest and the Apennine Fold Belt thrust from southwest to northeast. The South Apennine Basin was a Pliocene–Quaternary foreland basin. The Apennine and Alpine Orogenic Belts showed complex horseshoe-shaped structural morphology and controlled the Cenozoic tectonic evolution of the Molasse–Carpathian Basin, but had a weak influence on the remote structure of the European continent. The Dnepr–Nenets Craton Basin was long and narrow in shape, and had well-developed salt structure and high amplitude. It evolved from a rift basin and continues to sink to this day [34]. The Timan–Pechora

Foreland Basin was obviously asymmetric in the east–west profile. The west of it was a relatively flat platform, and the east was a foreland depression with a sharply thickened sedimentary cover. Basin subsidence was closely related to tectonic load. It went through the Paleozoic rift, with a passive continental margin and Permian foreland basin stages, forming a huge Permian system in the orogenic foreland. The Mesozoic evolution of the East Barents faulted in the Late Permian and Late Triassic and formed a large craton basin after long-term subsidence. The western part of the Yenisei–Khatanga Rift Basin was integrated with the western Siberian Giant Basin, a branch of the Triassic Rift System in the western Siberia, and intersected with the trigeminal rift system formed by mantle plume activity at the end of the Permian.

(2) **C5–C8: Meridional profile of Northeast China–Nansha–Indonesia Basin–Australia:** The Australia–Southeast Asia–China–Siberia meridional section passed through the Tethyan tectonic domain and the Central Asian tectonic domain. The main tectonic units passing from south to north included the Australian Plate, the Sumatra Subduction Zone, the eastern Indonesian continent, the South China Sea continent, the Southeast Asian Orogenic Belt, the Southeastern Orogenic Belt, the Xing'an Mongolian Orogenic Belt, the Yangtze continent, the Qinling Orogenic Belt, the North China continent, the Central Asian Orogenic Belt, and the Siberia continent. From south to north, it passed through the Bass Basin, the Eromanga Craton Basin, the Anderus Fault Basin, the Canning Craton Basin, the Australia Basin, the Java Basin, the South China Sea Basin, the Sichuan Basin, the Ordos Basin, the Hailar Basin, and the Siberian Basin.

Australia is now far away from the continental border and its northern and eastern parts are detached from the Pacific active area by the Papua and New Zealand microcontinents, which act as a barrier. Except for the collision with the Banda Arc in the northwest, the Australian continent is now an intracontinental tectonic unit with weaker tectonic activity. South Australia was in an extensional environment throughout the Late Mesozoic–Paleogene. As Australia was separated from the Antarctic continent, mantle magmatism had an impact and the internal stress field of the Australian continent changed. Southern Australia experienced lithospheric thinning and tectonic subsidence, resulting in the formation of rift basins. The initial rift between Australia and Antarctica began in the Late Jurassic and continued to develop into the Cretaceous. At Cenomanian

(95 Ma), Australia began to separate from the Antarctic continent and the cracking zone extended to the western part of Tasmania. The fault depression in South Australia mainly occurred during the Cretaceous–Paleogene development period and the Neogene tectonic reversal period, and generally had a fault-depression double-layer structure. The southernmost part of the Guppsland Basin was a faulted basin in southern Australia. From the early Cretaceous to the Jurassic, it was subjected to north–south stretching action and during the late Cretaceous it was in the post-fault thermal subsidence stage. From the Eocene to the present, the Champslan Basin suffered from compressional movement from northwest to the south. The Bass Basin was located in southern Australia, which was a semi-graben faulted basin within a stable block and was distributed in the northwest direction. The main fault development period in the Bath Basin was from the Early Cretaceous to the Early Eocene.

The Canning Basin was located in the central part of Western Australia and was an Early Ordovician–early Cretaceous cratonic basin. It experienced multiple sedimentary cycles since the Early Paleozoic and its southern depression had the largest thickness of 5,000 m. The Eromanga Craton Basin had an area of 120.56×10^4 km² and was the largest basin in Australia. It had a disc-like structure and basement asymmetry. Affected by short-term fault activities and thermal depressions, it settled slowly. The Cooper Basin and the Surat Basin were both part of the Eromanga large craton basin, and the Surat Basin was mainly from the Jurassic–Cretaceous. The Cooper Basin was a craton faulted basin from the Carboniferous to Triassic, which evolved into a part of Eromanga large craton basin after the Jurassic.

The Bonaparte Basin was a passive continental margin basin of the continental shelf of the Western Australia, which evolved from an internal craton basin in Gondwana, a rift basin in the continental cleavage period, and a passive continental margin basin in the continental drift period (Late Triassic to Early Cretaceous). During the late Miocene, affected by the collision between Australia's northwestern continental shelf and the Bandar Arc, salt diapirs accompanied by fault blocks and anticlines were formed. The current arc–continent collision in northwestern Australia is most obvious. Typical back-arc foreland basins developed in the arc–continent collision zone. The profile structure of the former defense basins was characterized by features such as the Papua Fold Belt in the Papua Basin and the Retroflexion Fold Belt in the Timor Basin outcropping the surface. The Southeast Asia–Australia region was located at the confluence of the Eurasia, the Indian–Australian continent, and the Pacific Plate. Australia subducted northward under the Sundaland and East Timor Arc. Timor Island was made up of a series of nappes, including ocean and continental material that rushed to the edge of the Australian continent. The foreland basins in Southeast Asia mainly consisted of the Timor and Papua Basins, which were transformed into foreland basins and passive marginal basins (the Bonaparte Basin) in northern and northwestern Australia.

The fore-arc basins in Southeast Asia were mainly located in the fore-arc positions of the volcanic arcs of Sumatra and Java. They were distributed along the side of the Indian Ocean and the main sediments were from island arc pyroclastic materials. Back-arc basins in Southeast Asia mainly included the Sumatra Basin and the Java Basins. They were formed in the Paleogene–Neogene Period and their structural evolution experienced rifting and reversal periods. They were also the most oil- and gas-rich basins in Southeast Asia (Liangqing Xue *et al.*, 2005). The South China Sea Basin was a marginal sea basin whose expansion began in the Oligocene, forming the current passive continental margin fault basin, nappe zone, subduction zone, sea basin, and other geological tectonic units. Compared to the Sundaland–Java Back-arc Basin, the basins in the South China Sea region were formed later. The South China Sea Basin's northern part was the passive continental margin basin (the Pearl River Mouth Basin), its southern part was the fore-arc basin (the Palawan Basin), and activity in the expansion center in the middle part had ceased. The South China Sea Basin went through the rift and subsidence stage during the Paleogene and the fault and subsidence stage during the late Neogene. It was an atypical passive continental margin basin and experienced transformations such as collision, compression, strike-slip, and oceanic crust subduction. Basins within the Asian continent were relatively stable in terms of structure and were weakly deformed, such as the Ordos Basin. The craton basin was affected by the tectonic activities at the edge of the plate, due to which it was uplifted, denuded, or shortened. Marginal depressions of craton basins developed as products of extension or deflection. The Mesozoic Hailar Rift Basin on the Xing'an Mongolian Orogenic Belt was relatively small in scale and had a short development history.

7.4. *South America–Africa Zonal Ultra-long Profile (D1–D4)*

The zonal long profile of the South America–Africa section (Figure 4) was approximately 7,500 km long and passed through the East Pacific

(a) Dl–D2: South American long profile

(b) D3–D4: Africa long profile

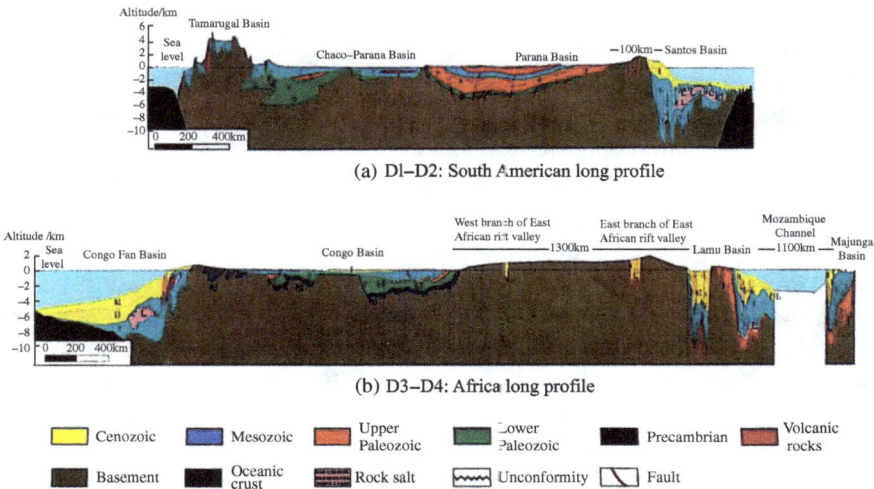

Figure 4.   South America–Africa zonal super-long profile (D1–D4).

continental subduction zone–South American continent–Atlantic Ocean–African continent from west to east. The profile crossed the Chaco–Barana Basin, the Santos Basin, the Lower Congo Basin, the Congo Basin, the East Africa Fault, the Lamu Basin, and the Majunga Basin from west to east. It showed the basin structure of the Gondwana continent since the Mesozoic cracking and the main features of the Paleozoic Craton Basins and the Mesozoic and Cenozoic Extensional Basins. The passive continental margin basin structure (depositional thickness and basin width) on both sides of the Atlantic Ocean was obviously asymmetric. The western margin of Africa was wider, while the Brazilian continental margin was narrower and the basement was strongly arched, recording the transfer of the late Pacific Ocean compressive deformation to the Atlantic region. The formation of the South Atlantic Ocean was the product of the early Cretaceous rifting of the Gondwana continent during the Late Jurassic Period. It was gradually split from south to north and began to drift between Africa and South America. Finally, it was completely separated. The opening time of the Atlantic Basin corresponded to the subduction of the Pacific Ocean, and the strong extension of the lithosphere was balanced by the subduction of the Pacific Ocean.

The South American profile showed the orderly parallel and continuous changes of trenches, fore-arc basins, island arcs, back-arc foreland basins, fault basins, craton basins, passive continental margin basins, and passive continental margin basins from west to east and was a typical profile reference area for classification of global basins. The foreland basin was located on the land-facing side of the Andes and was between the orogenic belt and the adjacent cratons. The Chaco Barana Basin, the Palana Paleozoic Craton Basin, and the Santos Passive Margin Basin were separated by continental arches. The continental arch was characterized by low amplitude and long-term development. The edge of the continental arch was continuously overlaid by the sedimentary strata of the adjacent basins, accompanied by the development of normal faults. The continental uplift became a group belt that separated or penetrated the craton basin from the foreland basin and passive continental margin basin in the same period. The Santos Basin was rich in oil and gas reserves in South America. Thick layers of evaporite and post-rift sediments were deposited in it. It experienced a syn-rift evolution stage during the Early Cretaceous Otterian–Early Apdi and a transitional evolution stage during the Apdi and the Albi–Holocene with passive continental margin evolution.

The African profile from west to east included the Lower Congo Passive Margin Basin, the Congo Craton Basin, the East African Rift Basin, the Lamu Basin, the Mozambique Strait, and the Western Madagascar Basin. The Lower Congo Passive Continental Margin Basin on the west coast of Africa underwent rift, transition, and drift stages since it was formed during the Cretaceous. Since the Oligocene, it developed the Congo submarine fan along the axis of the Lower Congo Basin. Due to the formation of thick Upper Cretaceous salt layers during the rift stage, the sedimentary strata of the basin developed extensional structural belts and compressional belts on the sea side due to gravity, and a thrust nappe structural belt appeared at the front end. The Congo Basin was a superimposed basin with Mesozoic and Cenozoic craton basins superimposed on Early Paleozoic rift basins. The East African Rift System was formed in the Paleogene–Quaternary Period and was also the world's largest onshore extensional fault zone. It was divided into the Eastern Ethiopia–Kenya Rift and the Western Tanganyika Rift. The East African Rift System was part of the African–Arabian Rift System and was caused by rising mantle plume activities. The African–Arabian Plate split and formed the Red Sea, the Gulf of Aden, and the Afar Trough trigeminal rift system. The former two further shaped a new ocean basin, and the latter formed a sky depression and extended south to the East African branch volcanic rocks and mountain ranges. It covered up to 2,000 m thick of Cenozoic basaltic lava and other volcanic lava and tuff. The western branch volcanic rocks were not well developed and formed into lakes. It was generally considered that the western branch volcanic rocks were relatively young. The formation of the eastern margin of the African continent was related to the breakup and drift of Madagascar in the Late Jurassic and it was formed at the same time as the Indian Ocean. The continental margin of East Africa was characterized by strike-slip and extension and thick Meso–Cenozoic deposits (thickness up to 6,000 m).

In terms of the Canadian Orogenic Belt, the Turpan–Hami Basin, the Junggar Basin, the Altai–Mongolian Orogenic Belt, the Siberia Basin, and the Laptev Sea Basin, the deposition centers of most basins were close to the side of the orogenic belt and were related to the development of foreland basins in later periods. Among them, the Paleozoic strata of the Junggar Basin were nearly 14 km long, the thickest section of the profile. The Tarim Basin and the Siberian Basin showed basement uplifts, which experienced subsidence periods at approximately 700 Ma and 1,600 Ma (multiple periods), respectively. On the profile, various types of sedimentary basins developed, including rift basins (the Laptev Sea Basin and the Santang Lake Basin), passive continental margin basins (the Mumbai Basin, the Qiangtang Basin, and the Eastern Siberian Sea Basin), and foreland basins (Ganges).

The basic features of the tectonic evolution of these sections included the development of uplifts and depressions in the geological historical period. This was true for both a continent and a certain area within a continent. For example, the Chinese continental area was a region of depression during the Cambrian–Middle Ordovician, which became a marine-covered area and began to accumulate marine deposits. In addition, there was similar evolution of other continents, including uplifts and depressions that led to land–sea evolution.

References

[1] Zengqian Liu. *Geological Tectonics, Formation and Evolution of Qinghai–Xizang Plateau*. Beijing: Geological Publishing House, 1990.
[2] Songnian Lu. *A Preliminary Study of Precambrian Geology in Northern Qinghai–Xizang Plateau*. Beijing: Geological Publishing House, 2002.
[3] Wenpu Ma. *Tectonic Properties of Snow Uplifts and its Transformation to Paleozoic Basin on Southeastern Margin of the Upper Yangtze River*. Beijing: Geological Publishing House, 1993.

[4]   Xingyuan Ma. *Compendium of Lithosphere Dynamics in China*. Beijing: Geological Publishing House, 1987.
[5]   Zijin Ma. *Global Meso-neozoic Tectonic Map (1:36,000,000)*. Beijing: Seismological Publishing House, 1996.
[6]   Peishi Miao, Xianqiang Zhou. *Global Tectonic System Map (1:2500)*. Beijing: Seismological Press, 2010.
[7]   Tingdong Li. *Asia and Europe Geological Map (1:5,000,000)*. Beijing: Geological Publishing House, 1996.
[8]   Zongjin Ma. *Global New Mesozoic Tectonic Map (1:36,000,000)*. Beijing: Seismological Press, 1996.
[9]   Zhenqiu Ren. *Global Changes*. Beijing: Science Press, 1990.
[10] Dianqing Sun, Peishi Miao. *Methods and Practices of Geomechanics*. Beijing: Geological Publishing House, 1999.
[11] Dianqing Sun, Qinghua Gao. Earth's rotation and global tectonics. Beijing: Geological Publishing House, 1980.
[12] Wegener [Austria], translated by Li Xudan. *Origin of Lands and Oceans*. Beijing: Commercial Press, 1964.
[13] Institute of Geomechanics, Chinese Academy of Geological Sciences. *Tectonic System Map of the People's Republic of China and its Adjacent Regions (1:2,500,000)*. Beijing: China Map Publishing House, 1984.
[14] Bosheng Zhang. *Crustal Mosaic Tectonics and its Basic Theory of Geology*. Xi'an: Shaanxi Science and Technology Press, 1983.
[15] Dawei Zhang, Dewu Qiao. *Investigation and Evaluation of National Oil and Gas Resource Strategic Selections*. Beijing: Geological Publishing House, 1993.
[16] Carroll, M. Supervision: Critical reflection for transformational learning (Part 2). *The Clinical Supervisor*, 2010 29(1): 1–19.
[17] Dengfa He, Guangming Zhai, Jun Kuang. Characteristics of the dynamics and evolution of the ancient uplift in the Junggar Basin. *Chinese Journal of Geology*, 2005 40(2): 248–261.
[18] Frolov, S.M., Basevich, V.Y., Aksenov, V.S. *et al.* Formation of nitrogen oxides in detonation waves. *Russian Journal of Physical Chemistry*, 2011 B(5): 661–663.
[19] Ci Bao, Xianjie Yang, Dengxiang Li. Geological tectonic characteristics and natural gas prospect Orediction in Sichuan Basin. *Natural Gas Industry*, 1985 7(4): 11–21.
[20] Yuqi Cheng. *Introduction to Regional Geology of China's Land*. Beijing: Geological Publishing House, 1994.
[21] Porter, S.C. *Late Quaternary Environments of the United States*. Minneapolis: University of Minnesota Press, 1982.
[22] Andrew D. Miall. *The Geology of Stratigraphic Sequences*. German: SpringerNature, 2010.

[23] Alejandro Escalona, Paul Mann. Three-dimensional structural architecture and evolution of the Eocene Pull-apart Basin, Central Maracaibo Basin, Venezuela. *Marine and Petroleum Geology*, 2003 20(2): 141–161.

[24] Jinxing Dai, Tingbin Wang. *Formation Conditions and Distribution Laws of China's Large and Medium-sized Natural Gas Fields*. Beijing: Geological Publishing House, 1997.

[25] Xianglin Shan. Determination of Qingbaikou system in Southern China. *Petroleum Geology & Experiment*, 1993 15(2): 146–159.

[26] Yuzhu Kang, Zongxiu Wang, Zhihong Kang, *et al*. *Research on Oil-controlling Effects of Tectonic Systems in Qaidam Basin*. Beijing: Geological Publishing House, 2010.

[27] Craig M. Bethke. *Geochemical Reaction Modeling: Concepts and Applications*. New York: Oxford University Press, 1996.

[28] Yuzhu Kang, Hongjun Sun, Zhihong Kang, *et al*. *China Paleozoic Marine Petroleum Geology*. Beijing: Geological Publishing House, 2011.

[29] Maclay, K. S., M. DeMaria, T. H. Vonder Haar. Tropical cyclone inner-core kinetic energy evolution. *Monthly Weather Review*, 2008 1(136): 4882–4898.

[30] Yuzhu Kang, Zongxiu Wang, Zhihong Kang, *et al*. *Research on Oil-controlling Effects of Tectonic Systems in Junggar-Tuha Basin*. Beijing: Geological Publishing House, 2011.

[31] A. R. Carroll, S. A. Graham, M.E. Smith. Walled sedimentary basins of China. *Geological Journal*, 2010 22(1): 17–32.

[32] Yuzhu Kang. *Unconventional Petroleum Geology in China*. Beijing: Geological Publishing House, 2015.

[33] Otofuji, Y. Large tectonic movement of the Japan Arc in late Cenozoic times inferred from paleomagnetism: Review and synthesis. *Island Arc*, 1996 5(3): 229–249.

[34] N.J. Kusznir, A. Kovkhuto, R.A. Stephenson. Syn-rift evolution of the Pripyat Trough: Constraints from structural and stratigraphic modelling. *Tectonophysics*, 1996 268(1): 221–236.

Chapter 5

Crustal Tectonic Deformation Styles

Abstract

Under the action of the geodynamics and a variety of geostresses, the global crust experienced eight structural styles in different geological periods. This chapter discusses tectonic deformation patterns of the crust, including the EW tectonic pattern, NS tectonic pattern, NE tectonic pattern, NNE tectonic pattern, NW tectonic pattern, epsilon tectonic pattern, S-shaped or reverse-S-shaped tectonic pattern, and twisted structure pattern. Finally, the evolutionary characteristics of tectonic patterns are introduced.

Keywords: Crust, structure, pattern, evolution.

After years of research, crustal tectonic deformation styles can be divided into eight types: the EW tectonic style, the NS tectonic style, the NE style, the NNE style, the NW style, the epsilon tectonic style, the S-shaped or reverse-S-shaped tectonic style, and the twisted structure style. But the main structures are the EW and NS tectonic styles (Figure 1).

This chapter describes the strengthening features of each structural system — phase, inheritance, difference, migration, and conversion — which show the complexity of each tectonic style.

This chapter [1] also points out that tectonic styles control the formation and evolution of landmasses of varying sizes, and each landmass, in turn, controls and influences the formation and evolution of tectonic

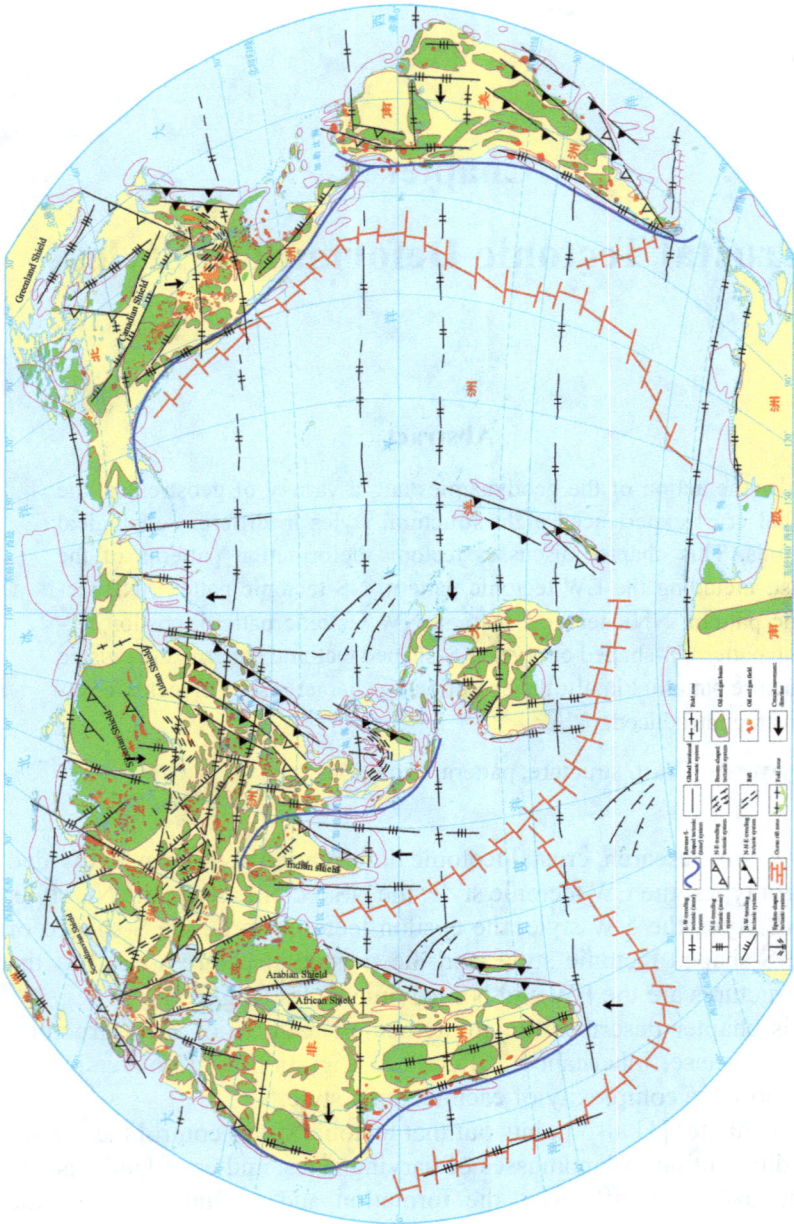

Figure 1.   Schematic diagram of distribution of major tectonic systems all over the world and oil/gas bearing basins (according to [2], supplemented and modified).

styles. Such interaction has created the current global tectonic styles as well as the changes and evolution of lands and seas.

The formation and the evolution of the tectonic system control the formation of sedimentary and prototype basins in various eras as well as the formation, transformation, and modelization of energy minerals and metal minerals all over the world. The tectonic style-controlled distribution of minerals was regular.

1. EW Tectonic Style

The EW tectonic style is a global tectonic style, which was distributed all over the world along a certain line of latitude. According to available data, belts of strong compression, magmatic activity, and metamorphic activity occurred at a rough latitude interval of $8°\sim10°$. Their main body was composed of various fold belts with an EW strike, compressive fault zones, and magmatic rocks. At the same time, they were diagonally intersected by torsional faults and vertically intersected by tensile faults; such belts generally contained or merged with some ancient blocks or rock blocks, with some EW troughs or basins present intermittently along the belts. As they were massive in scale, the depth of the crust was greatly affected. Intermediate-acid magmatic rocks as well as a wide range of mafic and ultramafic magmatic rocks appeared intermittently along the tectonic belts, distributed in phases. In addition, there were extensive and strong plastic-brittle deformations, often with large-scale tectonic dynamic metamorphic rock belts, such as low-temperature and high-pressure metamorphic rock belts, large-scale ductile shear belts, and migmatite-remelting granite belts [3]. Each tectonic belt formed a system of its own. The belts extended along a certain latitude, traversed continents and oceans, and were widely distributed across the world. Due to the interference of other structures and the slip or deflection of the block along the north–south direction, the current latitudes of each section were quite different and their directions were not all extending from east to west. The older the tectonic slip, the greater the deflection. Some sections of the giant tectonic system were often contained by newer east–west tectonic belts. Most of the giant tectonic systems in the world today were formed during the Late Paleozoic to Mesozoic and Cenozoic and some of them showed inheritance. The systems were wavy and sometimes sinusoidal in the direction of travel. Because some sections slipped along the EW fault zone (strike-slip) or the NS direction, several secondary derived torsional structural systems often occurred along the belt. The abovementioned factors complicated the belt's structural surface, so such a style is also called the giant EW complex structural belt [4]. The formation was restricted by the EW concord function zone produced by the rotation of the earth, so it showed distinctive orientation and positioning.

The EW tectonic zones have a globally distributed character. They are generally distributed along certain intervals of latitude, with a solid extruding tectonic zone occurring at intervals of 8°–10° in the middle and high latitude zones under normal circumstances. This was consistent with the critical latitude. Based on available data, the distribution of the EW tectonic style all over the world is shown in Figure 1.

2. SN Structural Style

The main body of the SN-trending tectonic style stretched in the SN direction. It was composed of a group of compressive structural belts or tension-fracturing structural belts running north–south and was accompanied by an NW-trending zone, an NE-trending twisting zone, and a near-east-trending extensional fracture. Each tectonic belt group formed a system of its own and was the product of the east–west squeezing and stretching of the earth's crust [5]. It was magnificent in scale and had corresponding deformation zones, magmatic activities, sedimentary formations, and sometimes tectonic dynamic metamorphic zones. The SN structural style affected the crust at different depths, some of which had a high impacted depth, reflecting obvious gravity and magnetic cascades, with basic and ultrabasic magmatic activities. In contrast, some had low impacted depths and insignificant magmatic activities, caused by the surface or shallow sliding of the crust. Some belts were generally only regional like some north–south structural belts in the south of China. However, regardless of the type or level of the north–south structural belts, their distribution in the upper crust was approximately equidistant and their mechanical properties were more territorial. The most striking thing was that in the Ural compressive north–south-trending tectonic system in central Eurasia and the north–south-trending mountain system between the western edge of the American continent and the east coast of the Pacific Ocean, the continental crust uptrending tectonic system varied in size and strength. The depths of the crust being impacted were different, but they were basically compression-oriented. The NS tectonic samples of

Europe, Africa, the Atlantic Ocean, and its coasts west of the Urals are dominated by the tensional rift, typical examples of which are the East African Rift and Sinking Belt, the Dead Sea in western Asia, the Jordan Valley and the Rhone Valley in west Europe, the Rhine Valley, the Scandinavian Rift, the Atlantic Rift, and so on [6]. This was probably the result of the kinematic mechanism of the Asian continent advancing faster while the American continent was slower in its crustal movement caused by the earth's rotation. It resulted in the east–west extrusion of the Asian continent and the east coast of Pacific Ocean, while Europe and the African continent, and the Atlantic Ocean were torn apart.

The north–south-trending tectonic system was one of the most basic and common tectonic systems in the crust and existed in both continents and oceans. Some sections were very magnificent, forming giant north–south tectonic styles (Figure 1).

3. NE Tectonic Style

The NE tectonic style was a multi-character tectonic style composed of a northeast–southwest-trending fold belt formed by the Caledonian and Indosinian Movements in the eastern crust of the Chinese continents. It was composed of deformation rock strata and deformation metamorphic belts formed by strong compression-shear direct torsion of rocks before the Late Triassic. NE Tectonic Style is mostly unconformably covered by the strata from the Late Triassic and their distribution direction was generally greater than 45°N eastward. This system was dominated by the NE-trending fold belt (or uplift and trough), accompanied by compressive and compression-torsion faults. In some areas, there were large-scale NE-trending metamorphic belts, granite belts, and volcanic rock belts. The northeast trending tectonic style was mainly formed and developed from the Paleozoic to the middle and late Triassic. It generally underwent two phases of deformation and metamorphism in the main curtains of the Caledonian and Indo-China movements. It was modelized at the end of the Middle Triassic to the early Late Triassic. It produced an important control effect on the Late Paleozoic and Triassic sediments and magmatic activities. Its deformation characteristics were dominated by plastic deformation, accompanied by low-, medium-, and high-pressure dynamic deformation metamorphic rock belts. Its strike was generally 45°~50° eastward [7]. Due to the cutting and division of the east–west structural belt and the differences in the basement structure of various regions, the development degree of each section was different and the deformation characteristics were different.

4. NNE Trending Structural Style

The NNE-trending structural style was composed of structural belts and sedimentary belts spreading north-north-east, typical of eastern China; it was developed in North America, South America, and the Pacific [8].

5. NW-trending Structural Style

The NW-trending structural style was mainly distributed in the northwest-northwestern tectonic system in western China and consisted of a series of parallel and roughly equidistant northwest-trending complex tectonic belts, which cut into into the east–west complex tectonic belt of the Tianshan Mountains at the southeast end of the eastern Junggar complex tectonic belt [9].

The NW-trending tectonic style of East Junggar in China included the Ertix Tectonic Belt, the Chukar Tectonic Zone, the Waugh Lake–Santanghu Subsidence Belt, the Beitashan Structural Belt, the Naomaohu Subsidence Belt, and the Kramari–Mochinwula Structure Belt.

6. Epsilon Structure Style

The epsilon tectonic style was composed of the following parts.

The front arc or frontal arc was an arc-shaped structure formed by a number of parallel compression zones and high-angle uplift zones, such as reverse faults, overburden faults, parallel schisms, and foliations. Within the northern hemisphere, this arc generally protruded to the south and only in a few cases to the west. For convenience of description, it can be divided into several parts. The middle or front part of the arc was called the top of the arc, and parts of the front arc that continued to extend back were called wings. In most cases, the top part of the arc exhibited the greatest curvature while the two wings showed only weak curvature [10]. Some fore-arcs differ little in the curvature of their tops and wings, forming a crescent together.

In each part of the curved extrusion belt, there were often tensile or torsional fractures approximately at right angles (these fractures were sometimes called transverse fractures). At the top of the arc, such faults were sometimes large in scale with the strata affected by them being deeper, causing the top of the arc to sink into a graben get covered by new sediments. Folds and thrust faults at the top of the arc sometimes formed similar arc-shaped structures under violent horizontal squeezing. Sometimes, the arc-shaped folds formed were not very wide and the area was not great [11]. The compression zones that made up the wings, including a long basin filled with the relatively new sediment of the uplift belt of ancient bedrock, were sometimes roughly parallel to each other and sometimes formed an echelon arrangement. The entire compression zone forming the two wings, including the folds, thrust faults, and elongated basins, tended to increase in number as it extended backward and its shape tended to spread out. The reflection arc was somewhere in the middle of the two wings of the front arc, and the arc began to show a tendency to reverse its bending direction; that is, the two wings tended to splay outward from there and moved in the opposite direction to the front of the front arc. The direction gradually bent and continued to stretch to the two end segments of the arc, forming two reflection arcs. When the front arc bulged to the south, the reflection arc bulged to the north. Accordingly, when the front arc bulged to the west, the reflection arc bulged to the east. Sometimes, the scale of the reflection arc was similar to that of the front arc; sometimes, the scale was smaller, the curvature was smaller, and the reflection arc only slightly curved outward [12]. The scale of the reflection arc were not crowded together in a number of narrower areas like the main part of the front arc, but were scattered over a wider area.

It should be pointed out that there was no boundary between the top, the two wings, and the two reflecting arcs of an epsilon structure. Together, they were slightly sinusoidal; that is, one side was S shaped and the other side was inverse S shaped. They were united at the vertices of the front arc to form a continuous, repeatedly curved compound structure belt. This did not mean that the various structural belts that comprised it were completely continuous from one end of the reflection arc to the other.

The ridge was in the depression area of the front arc, that is, in the middle of the area half surrounded by the front arc, and there was often a strong linear uplift and a compression zone. On rare occasions, this kind of uplifted extrusion zone underwent a process of settlement (quasi-trough-like settlement) before it uplifted. On some special occasions, it was still an open question whether it was possible for this uplift compression zone to become a trough zone after uplifting. The position of this uplift compression zone was roughly the same as the bilateral symmetry axis of the front arc, which was the ridge of the epsilon-type structure [13]. These complex compressive structural belts formed by several extruded belts were generally limited to a certain range, but sometimes were rather loose. The area, where the squeezing phenomenon was most severe, was mostly facing the apex of the front arc and its direction was roughly at right angles to the top of the front arc. On both sides of this strong squeeze zone, there were often weaker squeeze zones, which were further away from the central strong squeeze belts and appeared weak, and sometimes even disappeared. The compression belt that constituted the entire ridge was weakened as it got closer to the top of the arc and finally disappeared completely at a certain distance from the top of the arc. The abovementioned compression zones were composed of folds, uplifting surfaces, crushing zones, splitting surfaces, schistosities, and foliations. At right angles to the extrusion zone, there were often tensile fractures or normal faults [14].

What is described here is the normal form of the column, but whether it can also appear in the form of a wide uplift, that is, a small range of wide folds or multiple settlement zones (quasi-geochannels), is a question that needs to be further investigated. As there were often old rock formations in the area where the ridge was located, the compression zone forming the ridge was often compounded on the older compression belts. The compression directions of those older structures were of course not necessarily the same as those of the ridge of the epsilon structures. When the ridge part of the epsilon structure was formed, it had an uplift belt due to compression [15]. However, if this uplift zone was later subjected to tension in the direction that was at right angles to its axis, it was possible that, as mentioned above, large fractures could have occurred near or on both sides of it, forming a graben.

There were often horseshoe-shaped flat areas or areas with extremely weak folds between the ridge and the top arc and two wings of the front arc of the horseshoe-shaped betwixtolands. In places where the front curvature of the epsilon structures was not large, a vast and flat shield was often formed. This betwixtoland, which was a component of the epsilon structure, was composed of ancient folds, fractures, or other structural features that became rigid or possibly also composed of new folds, fractures, or other structural features running through the betwixtoland. All these old and new tectonic traces, of course, do not belong to the epsilon tectonic system. So, it must be clearly pointed out that their existence did not affect the stability of the horseshoe-shaped betwixtoland when it was formed. But, on occasions where the front curvature was very large, this horseshoe-shaped betwixtoland was still unavoidably affected by some relatively weak and minor axis folds. Some horseshoe-shaped betwixtolands, wholly or a part of them, were composed of ancient blocks that underwent folds or fractures up to the ground. There were also some horseshoe-shaped shield grounds, which were entirely or partially covered with a flat rock formation of certain thickness at the base of the ancient folds and fractures [16]. The former type of betwixtolands was sometimes called a platform and the latter was sometimes called a basin. However, from the perspective of the overall tectonic form of the epsilon structure, the basin on one side of the ridge and the ancient fractured and folded land exposed on the other side were more or less the same.

In addition to the main components of the abovementioned groups of epsilon structures, sometimes some second-level structural features occurred in concave areas of the reflection arc; but, in the sense of geomechanics, they were not necessarily secondary. In the recessed area of the reflecting arc, horizontal spiral structures of varying sizes often occurred. At the same time, within the range of the betwixtoland, especially in the middle of the horseshoe betwixtoland and the part not too far from the front arc, spiral structures sometimes occurred. The reflection arc depression area was generally a relatively stable area, which sometimes formed a basin or platform. But, sometimes, quite violent folds occurred in the middle area to form a reflection arc ridge. In that case, the relatively stable areas on both sides of it also became small horseshoe betwixtolands [17].

In front of the top of the forearc, sometimes granite bodies were exposed or buried in places that were not deep underground due to strong tension and cracking. At the top of the reflection arc, the same phenomenon sometimes occurred.

The ridge of the epsilon structure running north to south sometimes occurred in areas that were squeezed from east to west, including geosyncline and geoanticline that went north to south. Sometimes, because the epsilon structural ridge has already occurred, the subsequently rising tectonic movement took advantage of the momentum. In this way, the compound phenomenon of the epsilon structural ridge and the north–south structural belt was formed, which did not belong to the epsilon structural system. Such structures have been discovered frequently within the territory of China. If they appear behind the front arc of epsilon structure, especially in the middle behind the front arc, they will inevitably be mixed with components making up the ridge of the epsilon structure but they are not indistinguishable. As mentioned, the main difference between them is in the way they spread out or were distributed. The pure north–south structural belt often passed through the front arc of the epsilon structural systems, whereas the north–south structural belt belonging to the ridge can never pass through the front arc. Simple north–south structural belts were often strictly parallel to each other and spread over a wide range, whereas the various structural belts making up the epsilon ridge were densely packed in the middle area behind the front arc and often showed a trend of narrowing forward (that is, toward the top of the arc) and widening backward (that is, away from the top of the arc).

In a nutshell, the main components of the epsilon structures (except for a few cases where some were disturbed or damaged due to abnormal phenomena) involved the ridge as their axis, the two sides arranged approximately symmetrically, and the two wings as horns facing to each other, forming a whole with the abovementioned pattern. The structural elements making up its various components, such as folds and fractures, were also arranged according to a certain rule or interspersed with each other [18]. These arrangement rules played a certain role in controlling the distribution of minerals. Especially in the vicinity of the maximum curvature of the front arc and the reflection arc, sometimes, enrichment belts of deposits occurred. The above regularities must be addressed as a guide for our exploration program and construction design.

From the laws of these structural morphologies, we have discovered an important fact that the front arc of the epsilon structures in China generally bulged to the south and only a few epsilon structures had front arcs that bulged to the west. Judging from several epsilon structures that were identified in other parts of the northern hemisphere, they were also arranged according to the same rule. The direction of this epsilon structure is an amazing phenomenon in geological tectonics. It clearly points out that the origin of this type of tectonic style, like the east–west complex tectonic belt and the north–south tectonic belt, was inseparable from the current position of the earth's rotation axis.

According to the deformation image that considers the epsilon structure on the basis of theoretical analysis and simulation experiments, we have reason to treat the area involved in this type of structure as a flat slab beam.

The load borne by this flat beam was uniform and horizontal. The direction of the load was generally from high latitude to low latitude and in some cases from east to west. When this flat plate beam and the rock formation under it are fixed tightly only a short distance from the ends of the beam and are liable to slide or twist in other parts, it bends slightly in the direction in which the load is acting [19]. In the middle of the beam, that is, the place equivalent to the apex of the front arc, the bending was more obvious. At the same time, the epsilon structure line, especially the spreading, arrangement, and interpenetrating manner of the compressive and tensile structure lines, also reflected the shape of the main stress trajectory network in the flat beam. Places where the slab beam and the rock underneath it (which may be the so-called base layer, or may be deeper than the so-called base layer) were firmly anchored generally corresponded to the basement of the area where the concave reflecting arc was relatively stable.

The depth of the epsilon structures was still uncertain. But, generally speaking, the smaller the epsilon structures, the smaller the thickness of the rock layers that were affected. Meanwhile, the larger the scale, the greater the thickness of the rock layers that were affected. It is unclear if small and medium-sized structural systems belonged to this type. As far as the discovered epsilon structures are concerned, the smallest of them was more than 30 km long from the end of one reflection arc to the end of another reflection arc [20]. The distance from the apex of the outermost front arc to the farthest point of the ridge away from the front arc was more than 20 km. As for the scale of this type of structure, it is still uncertain to what extent it had reached.

7. S-shaped or Reverse-S-shaped Structure Style

An S-shaped structure is a type of twist structure. This type of tectonic style was generally large in scale, with a relatively complex shape and composition. There are various forms of composites of S-shaped constructions with other construction styles, and they commonly interfere with and utilize each other. According to characteristics of this tectonic style, it was generally divided into three parts — head, middle, and tail — but they were connected to each other as a whole and there was no division boundary. Generally speaking, the head was composed of a set of strongly curved or even hook-shaped folds and fault zones. The middle part was composed of a number of strong parallel folds and fault zones. Generally the trending is roughly close to north-south or west-east. A part of it was a slightly curved arc section, protruding to the west or east. The tail was also composed of parallel fold belts, generally also showing a curved shape but its bending direction was exactly the opposite direction of the head. In this way, the head, middle, and tail together formed a huge reverse-S-shaped structure [21]. The difference between it and the general reverse-S-shaped structure is that its head generally showed a strong twisting phenomenon. A part of the fault-fold belts of the head was often extremely curved, while the curvature of the tail was smoother than that of the head. The fault-fold belts around the head might have been loose and discontinuous. Therefore, there could have been several discontinuous, semi-circular twisted structures with unequal curvatures on the periphery of the head. Its central part generally rejoined or was mitered with the north–south tectonic style. In most cases, its tail was often composed of several arc-shaped fold belts extending from northwest to southeast to nearly east–west. There was often a stable plot in the center surrounded by these arc-shaped fold belts, which, contrary to the head, formed a sedimentary depression or vortex in the structure.

8. Twist Structure Style

The twist structure style was formed by a twisting motion centered on a twist shaft. There were three types of twisting shafts: upright, inclined, and horizontal. The twisting tectonic style generated by the rapid rotation of the inclined and horizontal shafts mostly occurred in the strata where the folds had moved and their scale was small, which can only be observed on the section and was not easy to see on the surface [22]. Upright twisting shafts are more common and embody three main forms: (1) a poorly developed broom-like structure; (2) a well-developed spiral structure; and (3) an extremely well-developed radiating concentric circular rotating structure. Regardless of the tectonic style, the common features are as follows: (1) The central part was composed of cylindrical or semi-circular rock blocks or blocks; (2) an arc-shaped torsional fracture developed, dividing the central part of the rock into a series of arc-shaped rock blocks; (3) this arc-shaped torsion fracture surface, whether it was tension-torsion or compression-torsion, surrounded a center and spread in concentric circles; and (4) the well-developed torsional tectonic style was accompanied by a series of radial torsional fracture surfaces, structural initiation zones, and rock blocks or blocks, including broom-like, geese-like, torsional, and radial tectonic styles.

The broom-like twisting structure style was generally composed of several arc-shaped structural features on the surface of the earth with one end converging and the other end spreading, which were half surrounded by a pillar or vortex (twist core) [23]. It was a poorly developed but widely distributed twisted tectonic style; that is, it was the product of the initial twisting stage. It can be seen often on different planes and sections and was widely developed in the crust.

The rotational surface of the broom structure can be a compressional-torsional fold bundle or fault zone or tensional-torsional fracture zone, or dyke swarm [24]. The former was called a compression-shear brush structure and the latter a tension-shear brush structure. Its twisting core can be a raised rock block, block or rock mass, or a sinking basin. Its mechanical properties were mainly determined by the mechanical properties of the cycle surface and the relative twisting direction. However, the properties had their own regularity; that is, inside the brush structure composed of tension, a tension-shear brush structure, and dike swarms, the internal rotation moved in the direction of spreading, whereas the external rotation moved in the direction of convergence. For brush structures composed of belts or compression-shear fractures, their internal rotation always moved in the convergent direction, while their external rotation always moved toward the spreading direction, which was exactly the opposite of the direction of movement of the cyclic surface of the tension-shear brush structure [25]. In short, if the arc-shaped structural surfaces making up the brush structure were tensile and torsional, it indicated that the rock surrounding the central part was twisted from the direction of spreading to the direction of convergence. If the structural surface was compression-shear, it implied that the twisted core of the rock in the center was twisted from the convergent direction to the spreading direction (Figure 2).

There are many examples of brush structures. The twist cores of general large and medium-sized brush structures were all vertical or nearly vertical, which reflected the plane twisting action in the region, while some brush structures with nearly horizontal axes were mostly small. This may be due to the limitation of the exposure depth. It was difficult to see or perceive the horizontal brush structures of some large and medium-sized rotating shafts. Detailed studies are needed to determine the features. In some large long-distance pushover structural belts, there may be such twisted structures.

(a) Tension torsion broom structure

(b) Compression torsion broom structure

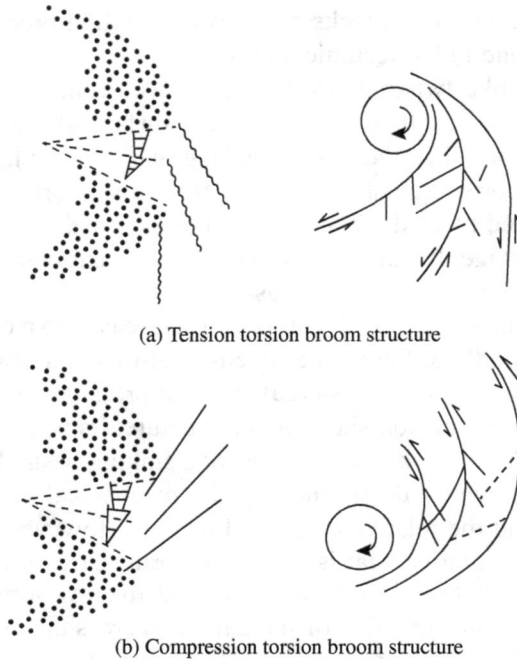

Figure 2.   Schematic diagram of the mechanical properties, twisting direction, and cause of the brush structure.

9. Evolution Characteristics of Tectonic Styles

Tectonic style is evolving and developing the main characteristics of the initial summary of five aspects: stage, inheritance, migration, difference, and conversion.

References

[1] Fuchuang Feng. *Natural Gas Geology in China*. Beijing: Geological Publishing House, 1995.

[2] Guoyu Li, Zhijun Jin. *World Atlas of Oil and Gas Basins*. Beijing: Petroleum Industry Press, 2005.

[3] Shicong Guan. *Sedimentary Facies and Oil and Gas in China's Sea-land Transition Area*. Beijing: Science Press, 1984.

[4] Zhengwu Guo. *Research on Formation and Evolution of Sichuan Basin*. Beijing: Geological Publishing House, 1996.

[5] Jianyi Hu. *Geological Basis and Oil and Gas Enrichment in Bohai Bay Basin*. Beijing: Petroleum Industry Press, 1990.

[6] Jiqing Huang. *Research on Tectonic Characteristics of China*. Beijing: Geological Publishing House, 1984.

[7] Qinghuan Jin. *Geology and Oil and Gas Resources in South China Sea*. Beijing: Geological Publishing House, 1988.

[8] Yuzhu Kang, Zhenwei Gan, Zhihong Kang, *et al. Oil and Gas Distribution Law and Exploration Experiences in China's Main Basins*. Urumqi: Xinjiang Science and Technology Press, 2004.

[9] Yuzhu Kang. *Distribution Law and Development Strategies of Global Oil and Gas Sources*. Beijing: Geological Publishing House, 2016.

[10] Yuzhu Kang. *Introduction to Global Tectonic System*. Beijing: Geological Publishing House, 2018.

[11] Xiguang Deng, Xiangshen Zheng, Xiaohan Liu. Discovery of gravel-bearing mudstone layers in Livingston Island in Antarctica and its geological significance. *Polar Research*, 1999 11(3): 169–178.

[12] Kang Zhang, Zongying Zhou, Qingfan Zhou. *China Petroleum and Natural Gas Development Strategies*. Beijing: Petroleum Industry Press, 2002.

[13] Qi Zhang, Xiaomei Fang, Linhua Guan. *World Oil Producing Countries (North America and Europe)*. Beijing: China National Petroleum Corporation, 1998.

[14] China Petroleum Institute of Economics and Technology. *2006 Overseas Oil and Gas Investment Environmental Monitoring and Analysis Report*. Beijing: China Petroleum Institute of Economics and Technology, 2006.

[15] Qiashun Wang, Dagang Zhu, Chengyun Xiong, *et al. Methods and Practices of Geomechanics — Introduction to the Second Tectonic Systems*. Beijing: Geological Publishing House, 1999.

[16] Xuchang Xiao. *Geotectonics of Northern Xinjiang and Adjacent Regions*. Beijing: Geological Publishing House, 1992.

[17] Compilation Group of Regional Stratigraphic Table of Xinjiang Uygur Autonomous Region. *Xinjiang Autonomous District Book of Regional Stratigraphy Table in Northwest China*. Beijing: Geological Publishing House, 1981.

[18] Guangming Zhai. *China Petroleum Geology*. Beijing: Petroleum Industry Press, 1996.

[19] Fuli Zhang. *Natural Gas Geology in Ordos Basin*. Beijing: Geological Publishing House, 1994.

[20] Guojun Zhang, Jun Kuang. Petroleum geological characteristics and oil prospects in Hinterland of Junggar Basin. *Xinjiang Petroleum Geology*, 1993 14(3): 5–12.

[21] Yuchang Zhang. *Prototype Analysis of Petroliferous Basins in China*. Nanjing: Nanjing University Press, 1997.

[22] Zhiwu Zhou. *Geological Tectonic Characteristics and Petroleum-bearing Properties of East China Sea.* Beijing: Petroleum Industry Press, 1990.

[23] Xia Zhu. *Tectonics and Evolution of Mesozoic and Cenozoic Basins in China.* Beijing: Science Press, 1983.

[24] Jishun Ren. *1:500000 Geotectonic Map and Descriptions of China and its Neighboring Regions — China's Geo-tectonics from Global Perspective.* Beijing: Geological Publishing, 1999.

[25] Xiaoguang Tong, Lirong Dou, Zuoji Tian, *et al. Research on China's Transnational Oil and Gas Exploration and Development Strategies in Early 21st Century.* Beijing: Petroleum Industry Press, 2003.

Chapter 6

Features of Crustal Uplifts and Depressions All Over the World

Abstract

This chapter discusses evolutionary features of crustal uplifts and depressions all over the world. Global patterns can be explained as uplifts and depressions. Lands appear as a result of uplifting of the earth, whereas oceans are formed as a result of sinking. Uplifts on continents formed mountains, whereas depressions resulted in lakes or swamps. The crust evolved in this way throughout the history of the earth. There were no continental drifts or plate movements. The global crust can only be formed by the evolution of uplifts and depressions as well as that of seas and lands.

Keywords: Global, crust, uplifts and depressions, evolution.

Based on the generation reservoir stress, characteristics of crustal movements, tectonic movements, large earthquakes, seawater transgression–regression, volcanic eruption, and other phenomena in the crust in the recent one million years, the author maintains that no landmass on the globe can drift [1]. After analysis in the early 1920s, Wegener put forward the hypothesis of continental drift, which has been circulated for a while. Nowadays, there are still some people who believe that continents can drift. This is a fallacy without facts.

The crust can only cause deep and large fault activities because of horizontal compressive force, resulting in ascending uplifts, sinking depressions, and strike-slips subduction and uplifts. In particular, when the tensional deep fractures are pulled apart, the magma in the mantle will invade or erupt and will seal the fractures like a gel. Continental landmasses changed a lot during geological historical periods. This change was not mainly due to the large-distance displacement of the continental landmasses themselves nor because of the continent drifting, but due to the land–sea changes caused by the uplift and depression of continents. For example, during the 2011 Fukushima Earthquake in Japan, the large-scale tsunami caused land sinking and depression. The land area of the Japanese Islands was reduced due to submersion under ocean water, and not because of the drift of the islands. It is inferred accordingly that changes and displacements of the landmasses in the long geological history of earth were mainly the result of the changes of the land and sea caused by the evolution of the crustal uplift and depression [2].

Of course, deep and large faults caused by *in-situ* stress have the following characteristics: (1) squeezability can make a part of the landmasses dive under another block; (2) squeeze can create mountains (uplifts); and (3) strike-slip faults can make two fractured discs slide for several kilometers. But, this relative motion is a local motion for the earth as a whole. In a certain historical period, ocean changes had the greatest impact on landmass changes, such as changes of various continents in different periods. Various types of tectonic systems formed in the crust due to this, with their own characteristics.

There were three main forms of crustal movements: extensional depressions, squeeze uplifts, and strike-slip depressions and uplifts. These three forms existed correspondingly. If there was an uplift, there must be a depression.

If strong squeezing occurs in adjacent uplift areas, it will inevitably cause large-scale depressions in adjacent areas. According to both theoretical analysis and actual process investigations, fold orogenic belts must have been subjected to strong horizontal compression, whereas large-area subsidence can provide strong horizontal compression [3]. In addition, due to the large-area subsidence process, heat energy, gas, liquid, and magma from the deeper parts were forced to move horizontally in the form of an arc parabola with the arc top downward, and to move, invade, and rise to the adjacent rising place and further zones, thus forming a strong lifting action of opposite convection and collision in the deeper part of the uplift belt. This is another basic and important feature of crustal movement dominated by subsidence.

Like splitting wood, the settlement process will produce lateral pressure to generate a strong squeeze. Given the conditions of relatively low elevation of the adjacent ascending zones, as well as the soft and plastic surface rock and soil, the tiled state, and other characteristics, under the action of strong extrusion movement during settlement, it is easy to produce features like strong deformations, folds, and nappes. Various phenomena that occurred in the early stage of the geotrough evolution can best illustrate this point [4]. According to geophysical data, the edge of the Atlantic Ocean, the Indian Ocean, and the Arctic Ocean are unconformably superimposed on the adjacent continents, which fully demonstrates the huge-area subsidence process of ocean basins and sea basins, resulting in strong horizontal dynamic movements.

High-pressure and low-temperature metamorphic rocks are found on the land walls of trenches. The rocks are strongly deformed, resulting in low-angle thrusts and folds of the nappe. This also fully shows that the large-scale subsidence process of ocean basins and trenches resulted in huge horizontal dynamic movements.

It is common to see alternating and repeated occurrences of meganticlines and megasyncline in regional geology. Among them, the strata of meganticlines are strongly folded, ascended relatively and with some strata missed. Igneous rocks are common among them and fractures are well developed [5]. However, in the megasynclines, the surface rock strata are gentle and often have more strata of several ages than the adjacent compound megasynclines. How did the alternating meganticlines and megasyncline come into being? Where did the horizontal opposing force forming meganticlines come from and why? The uplift theory believes that the formation of the mid-ocean ridge zone was the result of strong squeezing by the horizontal lateral forces generated by the subsidence and depression process from both sides. At the same time, due to the subsidence process of megasynclines, heat energy, gas, liquid, and magma in the deeper part of the belt were forced to transfer, invade, and rise in the reciprocating megasynclines, which formed strong uplift of opposite convection and collision. When a large fracture occurs in the middle section of a meganticlines, and causes an earthquake, with magma intrusion and volcanic eruption, energy in the deep part of the belt is greatly consumed. After the rocks are broken by fracture, their compressive strength and bearing capacity are greatly weakened. In this way, a graben is formed along the fault or a lower subsidence similar to a small basin is formed [6, 7]. Through this natural adjustment, the dynamic torque and resistance torque in the region can maintain a new balance and a relatively stable state.

Based on analysis of the distribution of sedimentary rocks, tectonic movements, and denudation at different times all over the world, it is concluded that the Sinian had a larger land area, while the Early Paleozoic–Cambrian had the smallest land area and the largest ocean area. Moreover, the Ordovician had more land area compared to the Cambrian, the Silurian had the largest land area in the Paleozoic Era, the Late Paleozoic–Devonian had a smaller land area, and the Permian had the largest land area. Since the Mesozoic, the land area has been increasing and the sea water area has been shrinking [8]. In the Paleogene, the embryonic form of the current continents was formed.

Land–sea changes were definitely not due to the drift of the continents, but the following two factors:

> First, the squeezing effect of the crustal stress caused different parts of the crust to rise and other parts to sink. The sea water flowed to the subsidence zone and the uplifts became lands. Crustal uplift and subsidence were constantly and unevenly carried out, so were the seawater transgression and regression. In addition, factors affecting seawater transgression and regression all over the world included the reduction of seawater area during the global ice age and the increase of land area, such as in the Late Sinian, the Late Ordovician–Early Silurian, the Late Carboniferous–Early Permian, and the Early Quaternary [9].
>
> Second, the ground stress caused large-scale fractures in the earth's crust in multiple directions, which promoted relative movement of the earth's crust, such as relative lifting or relative translation along the faults, and promoted relative uplift and settlement or translational movements of the land.

References

[1] Desheng Li, Yonggeng Yao. Geological characteristics of petroliferous basins in western China. *Petroleum Exploration and Development*, 1991 18(2): 11.
[2] Dongxu Li. *Introduction to Geomechanics*. Beijing: Geological Publishing House, 1986.

[3] Yinsheng Ma. *Neotectonic Activities and Geological Disaster Risk Assessment in Upper Reaches of Yellow River*. Beijing: Geological Publishing House, 2003.

[4] Bureau of Geology and Mineral Resources of Inner Mongolia Autonomous Region. *Regional Geology of Inner Mongolia Autonomous Region*. Beijing: Geological Publishing House, 1991.

[5] Chongzhi Ning. *Discussion on Arc Top and Spine Structures of Huaiyang Epsilon-shaped Tectonic System based on Geological Tectonic Outline of Dabie Mountain in Ele Region*. Beijing: Geological Publishing House, 1959.

[6] Huang, P. H., Fu, R. S. The mantle convection pattern and force source mechanism of recent tectonic movement in China. *Physics of the Earth & Planetary Interiors*, 1982 28(3): 260–268.

[7] Zhang Yang, Shumo Ge. Preliminary study of the fracture zone by 1931 Fuyun Earthquake and the features of neotectonic movement. *Seismology and Geology*, 1980 3(4): 33–39, 84.

[8] Barosh, P. J. Neotectonic movement, earthquakes and stress state in the eastern United States. *Tectonophysics*, 1986 132(1–3): 117–152.

[9] Bian, J., Tang, J., Lin, N. Relationship between saline–alkali soil formation and neotectonic movement in Songnen Plain, China. *Environmental Geology (Berlin)*, 2008 55(7): 1421–1429.

Chapter 7

Evidence of Sea–Land Changes

Abstract

This chapter touches on evidence of land–sea changes, including local and global land–sea changes. Crustal fault activities, volcanic eruptions, and seismic activities are all manifestations of the imbalanced effects of local geostresses, which however are local phenomena of separation movements and thus cannot cause the separation of the entire crust at all. Therefore, the earth's crust will always be a complete and indivisible whole as it was in the past.

Keywords: Transformation between seas and continents, global, whole, crust.

1. Local Sea–Land Changes

According to Belousov's comprehensive geological data on the seabed structure, and continental margins, local sea–land changes were intensified toward oceans.

According to Stovaz's data, subsidence depressions were common on the east and west coasts of the American continent, and a large part of mountainous area on the west coast is submerged in ocean water today. In many places near the coast, it was gradually discovered that the land was submerged in the sea. Many places, such as the Bering Sea, the Sea of Okhotsk, the Sea of Japan, the Yellow Sea, the Bohai Sea, the East China Sea, the South China Sea, the Bay of Bengal, and the Arabian Sea, which were still land in the Tertiary Period, 10 million years ago, have become seas today [1].

Recently, a continent that was approximately the size of Argentina and had subsided 1.6 million years ago was discovered in Mount Carroll in Antarctica, a fact confirmed by dinosaur fossils and coal deposits obtained from the seabed.

Almost the entire south coast of the Baltic continent was sinking slowly. Similarly, the southern coast of the North Sea, the coast of English Channel, the vicinity of Sukhumi by the Black Sea, part of the coast of the Australian continent, the coast of Siberia near the Arctic Ocean, and many places near the mouth of the Congo River in Africa are all in a state of slow subsidence today.

At the end of the Paleozoic, during the early Mesozoic, and even the Old Tertiary, large areas of the Mediterranean were uplifted into land. Later, due to the stretching effect, a large area of depression occurred in the adjacent coast, which has now become an inland sea and is still expanding [2].

In the 1960s, thousands of "smokestacks" were discovered at the bottom of the Red Sea. Due to the eruption of submarine volcanoes, the coast of the Red Sea slowly subsided and sank. The sea surface continued to expand and the Red Sea became the sea with the highest water temperature in the world. Later, "smokestacks" were also found on the seabed in western Mexico. This kind of submarine volcanism put the Mexican coast and other places in a state of slow subsidence. With the development of geology and archaeology, the ancient continent of Gondwana was discovered in the southern hemisphere. Now, most of the land has sunk into the Indian Ocean and the Atlantic Ocean.

Subsidence and areas of depression were gradually discovered in the interior of the continental shelf. For example, the front mountains of the Alps, Bavaria, areas near lakes, and the shores of Lake Michigan in North America are all slowly sinking today. The southern coast of England is also sinking today. The subsidence rate of the Dutch coast is about 3 mm [3] per year. The Gulf of Naples, Italy, sinks approximately 7 mm a year.

The alluvial plain of Hebei in China is in a state of slow subsidence, whereas the Taihang Mountain is relatively uplifted. About 1 million years ago, the Hebei Plain was submerged by sea water and Beijing was a bay at that time. Some parts of the Hebei Alluvial Plain have subsided as much as 1,000 m. If calculated by 1 million years, the subsidence rate is about 1mm per year. However, the Haihe River, and especially the Yellow River, continuously deposited a large amount of sediment onto the seabed, compensated the height lost by subsidence, and regained the area occupied by sea water, causing the alluvial plain to continuously expand into the sea. (In fact, the Yellow River today enters the sea in Shandong Province, but once entered the sea from the present Hebei Province and Tianjin City; according to observations, the river-bed in the lower reaches of the Yellow River rises by an average of 10 cm every year.)

According to the article "The Latest Monitoring Results Show that Beijing is Sinking" published in 1999, aside from land subsidence in the eastern suburbs of Beijing that was under control at that time, three obvious land subsidence areas appeared in the surrounding regions. The settlement range in 10 years was 337~385 mm and the average annual settlement was more than 30 mm [4]. The reasons contributing to this great subsidence were not only the excessive exploitation of groundwater but also the slow subsidence of the entire Hebei Alluvial Plain. The difference between the present and the past lies in our ability to effectively control land subsidence. Controlling subsidence can be considered a victory. A stele in Changli County, Hebei, indicates that the town was 2.5 km away but is now only 1 km away from the seaside, indicating that the coast is slowly sinking.

A strong earthquake occurred in Xingtai, Hebei, in 1966. Results of measurements by the National Surveying and Mapping Administration showed that the fracture subsidence zone produced by the earthquake matched the range of the extreme earthquake zone. The fault subsidence zone is to the NNE, with a subsidence range of 315–714 mm and a neighboring ascending range of 40–72 mm. In the early morning of July 28, 1976, a 7.8-magnitude earthquake occurred in Tangshan, Hebei. The subsidence in the extreme earthquake zone was great and the area almost fell into ruins with 242,000 people losing their lives [5].

In 1885, a sea earthquake hit the Adriatic Sea in Italy, causing the seabed depth to change from 200–3,000 m. During the eruption of Mount Pelée on the island of Martinique, the depth of the neighboring parts of the sea increased by several hundred meters. After three or four volcanic eruptions, the Krakatoa Island of Japan sank into the sea.

Crustal movement causes crustal uplift and depression to proceed without interruption. For example, sedimentary rock layers of the Appalachian Mountains in North America had a thickness of 10–12 km.

And the sedimentary rock layer of Luopu Mountain was approximately 18 km thick [6]. According to field and indoor research conducted by a Chinese expedition team from 1966–1968, the thickness of knotted crystalline rock in the Himalayas was more than 20 km.

Looking back at history, during the Archean, a generally strong folded crystalline rock shell was formed in China's land area as the earth had just begun to cool and solidify, and at that time, the crustal movements and magmatic activities were intense and volcanic eruptions were frequent. During the Proterozoic, in the second half of the Sinian period, almost the whole of China's land area was submerged by due to subsidence and depression. This area ere was submerged by seawater more than once in the Paleozoic, such as in the Upper Cambrian, the Ordovician, and the Middle Carboniferous–Permian periods. When it came to the Mesozoic and Cenozoic, due to the sinking of the Pacific Ocean on a large scale, China gradually rose. Moreover, the terrain was high in the west and low in the east, and the seawater gradually withdrew. Since the Neogene, the "roof of the world" — the Qinghai–Xizang Plateau and the Himalayan mountain system — has been formed, with a maximum height of more than 8 km [7].

In the vast oceans, some new islands that were slowly rising were not always rising, and have experienced repeated uplifts and depressions. For example, near the western islands in Tonga in the Pacific Ocean, there was an island called "Jack-in-the-box." In the 30 years from 1904 to 1938, there were a total of six depressions and uplifts of this island. The Tongans once revealed that the Lord of the Sea gave them a new life. But, the island then sank. Later, some geologists wearing diving suits, goggles, and oxygen tanks drilled into the bottom of the sea at a depth of 33 m where they discovered that submarine volcanoes were still erupting. In the waters of the Mediterranean Sea near Sicily, people saw the bottom of the sea boiling like a pot on July 10, 1831, and heard a muffled sound after which a tall and dazzling plume of red smoke rose from the water [8]. A week later, a small island a few meters high appeared there. Another week later, it was approximately 20 m above the water surface. By August 4, 1831, the small island had grown to a height of approximately 60 m and the circumference of the island was approximately 1 mile (1.852 km). But then the island disappeared due to sinking. In December of the same year, this small island emerged from the sea again overnight. For many years, it went up and down, and repeated this process several times. It was not until 1950, when diplomats of several countries were arguing about its sovereignty, that it suddenly sank and disappeared, leaving surging water in the area.

According to geodetic surveys, the triangular points on the northern edge of the Bavarian Alps moved 0.25–1 m northeast toward Munich from 1801–1805. In the same way, it was proved that Lake Geneva, along with the Alps around it, was moving north.

According to repeated measurements made by Chinese scientists using common length baseline radio technology, the distance between Shanghai and Japan, the United States, and Australia was shortened at a rate of 2–8 cm per year. The earth is dominated by contraction and sub-sidence. The Earth is dominated by contraction and subsidence, and it is thought that the evolution of the Pacific Ocean floor has entered a return phase, with consequent ocean-floor earthquakes, magma intrusion, and volcanic eruptions. Shallow basaltic magma gushed out in large quanti-ties but no substance erupted in the air, showing gentle characteristics [9]. In this way, the adjacent ocean basins and many places along the coast of the Pacific Ocean showed a large-scale subsidence process and a certain degree of horizontal opposing movement. The distance between Shanghai and Japan, the United States, and Australia was shortened at a rate of 2∼8 cm per year, which was a natural consequence.

On the coast of California in North America, all survey points north of Golden Gate, accurately set by geodetic methods, moved northward at an average annual rate of approximately 5.2 cm during the 40 years from 1868 to 1906. In 1906, an 8.3-magnitude earthquake hit San Francisco. All the measuring points suddenly moved southward by 1–2 m and then started to creep north slowly again. This was connected to the horizontal dynamic moment caused by the large-scale settlement of the adjacent Pacific Ocean bottom and California Bay. When a major earthquake hit San Francisco, it suddenly moved south because the enormous energy in the deep part of the zone was consumed. After this natural adjustment, the acting direction of forces always moved toward the direction of least resistance and then resumed the original slow movement northward. This is a natural occurrence.

The Hawaiian Islands moved horizontally westward at an average speed of about 5 cm per year, which was due to the results of the horizon-tal dynamic moment produced in the subsidence process of the Pacific Ocean basin.

2. Global Sea and Land Changes

2.1. *The Precambrian*

In the Late Proterozoic, all the continents began to converge with each other at about 1,100–1,000 Ma to form the supercontinent Rodinia. The Rodinia Supercontinent began to split at 750 Ma when the earth entered a period of ice age. The temperature all over the world dropped sharply and the largest ice age in geological history began.

Rodinia split into three landmasses at 750 Ma: Proto-Laurasia, Congo, and Proto-Gondwana. The ancient ocean of Panthalassic began to form, proto-Laurasia further rifted and turned to Antarctica, and proto-Gondwana rotated counterclockwise [10]. According to the available paleomagnetic and geological data, Scotese *et al.* (1997) reshaped the paleogeographic map at 650 Ma (Figure 1).

During the period of 600–550 Ma, due to Pan-African fold orogenic activities, these three continents once again converged into a theoretically new supercontinent — the Pannotia Supercontinent. Most of the Pannotian continent was located in the polar region. Evidence shows that during this era much of the earth was covered by large areas of glaciers, much larger than any other period in the geological era.

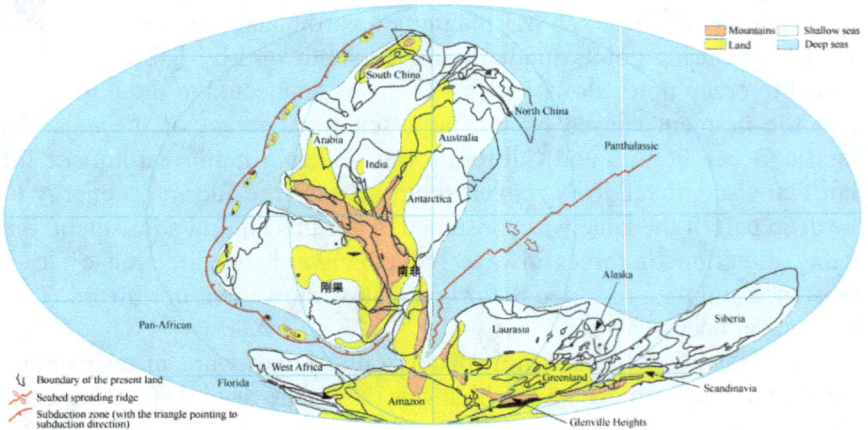

Figure 1.   Global plate reshaping map during the Late Proterozoic (650 Ma) (according to Scottese [11], slightly modified).

2.2. *The Early Paleozoic*

At 540 Ma (the Early Cambrian), the Pannotian Supercontinent that was formed at the end of the Late Proterozoic split into Laurasia (North America), Baltic (Northern Europe), Siberia, and Gondwana. The Iapetus Ocean expanded between several ancient continents. Among them, the Laurasian continent was located near the equator and the Gondwanan continent merged on the non-fold belt and was formed near the Antarctic Ocean (Figure 2). During the Cambrian period, the continents were covered by shallow seas and for the first time, large numbers of creatures with hard shells appeared.

During the Ordovician period, the continents of Laurentia, Baltica, Siberia, and Gondwana were separated by ancient oceans. The Iapetus Ocean separated the Baltic and Siberian continents. The original Pro-Tethys separated the Gondwanan, Baltic, and Siberian continents. A vast ancient ocean covered most of the northern hemisphere (Figure 3). In Gondwana, limestone and evaporite were deposited in the warm equatorial region, whereas glacial clastic deposits were made in the southern region. During the Early Ordovician (approximately 480 Ma), the Avalonian microcontinent was separated from the Gondwanan continent and migrated to the Laurentian continent [12]. At the end of the

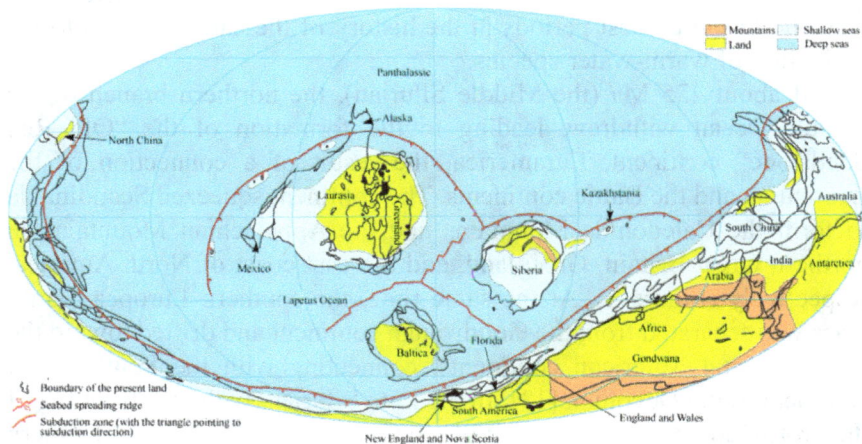

Figure 2. Global plate reshaping map during the Cambrian period (514 Ma) (according to Scotese [11], slightly modified).

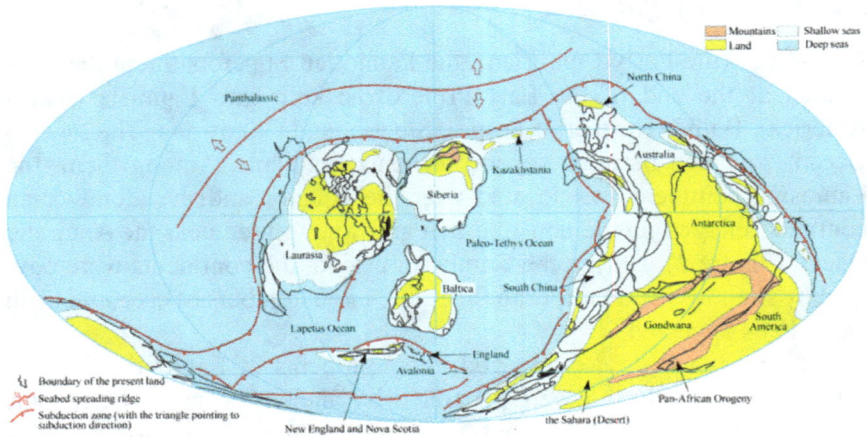

Figure 3.   Global plate remodeling map during the late Ordovician period (458 Ma). (according to Scotese [11], slightly modified).

Ordovician, the Laurentian, Baltic, and Avalonian microcontinents began to converge into the Euramerica, the Iapetus Ocean began to close, which also prompted the formation of the Appalachians, and the Gondwanan continent to migrate slowly toward Antarctica. At the end of the Ordovician, the earth's climate began to enter the Great Ice Age, which was one of the coldest periods in the history of the earth, leading to the extinction of warm-water species.

At about 425 Ma (the Middle Silurian), the northern branch of the Iapetus Ocean withdrew, leading to the formation of the "Old Red Sandstone" continent, Euramerica, the result of a connection of the Laurentian and the Baltic continents. The continent squeezed Scandinavia to form the Caledonide Mountains and the Appalachian Mountains in northern Great Britain, Greenland, and the east coast of North America, respectively (Figure 4). At the same time, the southern European land-mass was separated from the Gondwanan continent and drifted toward the European and American continents, connecting with the South Baltic continent in the Devonian [13]. In the Late Silurian, with the shrinking of the Proto-Tethys Ocean, a new Proto-Tethys Ocean formed at its southern end. The South Chinese and North Chinese continents were separated from Gondwana and moved northward.

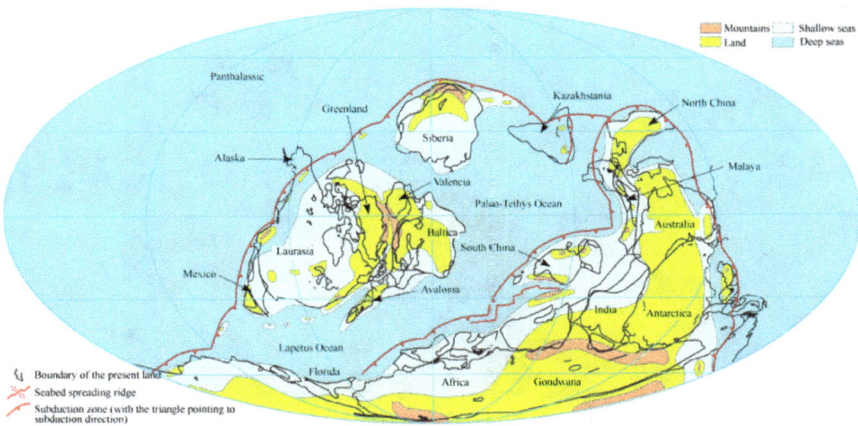

Figure 4.   Global plate remodeling map during the Silurian period (425 Ma) (according to Scotese [11], slightly modified).

2.3. *Late Paleozoic*

During the Devonian, the Gondwanan continent began to migrate to the European and American continents, and the ocean formed in the Early Paleozoic began to withdraw to form the primitive pan-continent — Pre-Pangea (Figure 5). Plants began to appear in large numbers. For the first time, forests appeared in the equatorial region. Abundant coal was formed in today's northern Canada, northern Greenland, Scandinavia, and other ancient tropical swamps.

In the Early Carboniferous, the European and American continents joined the Gondwanan continent, whereas the Paleozoic oceans retreated and the Appalachian uplift became a mountain range, as did the Williscan uplift (Variscan Mountains) (Figure 6).

The South American continent migrated to the south of the European and North American continents, whereas eastern Gondwana migrated from the equator to the Antarctic Ocean. Southern China and Northern China were independent continents during this period. At the same time, Kazakhstania was connected to the Siberian continent [14]. An ice cap was formed in Antarctica when quadruped ridge animals began to evolve in coal swamps near the equator.

In the Late Carboniferous, the western part of the Kazakhstania continental area was connected to the Baltic continent, and the ocean

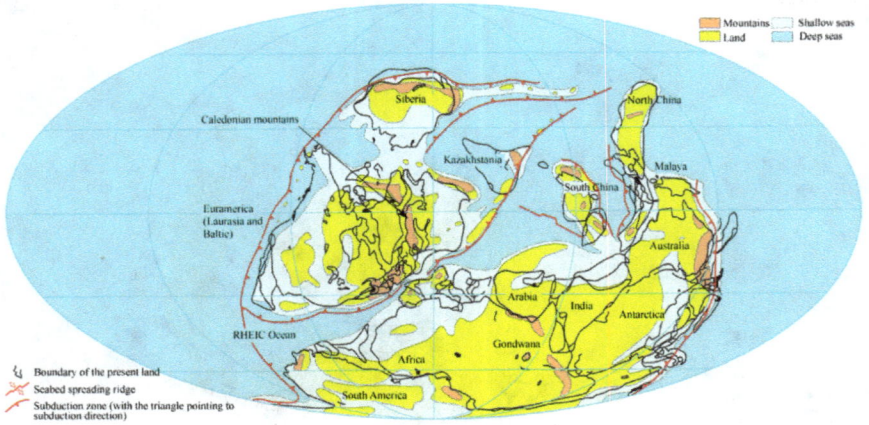

Figure 5.   Global plate remodeling map during the Early Devonian (390 Ma) (according to Scotese [11], slightly modified).

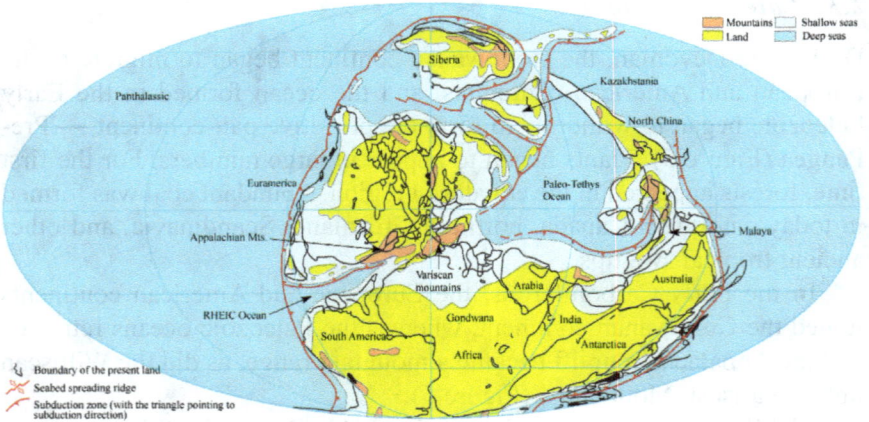

Figure 6.　Global plate remodeling map during the Early Carboniferous (356 Ma) (according to Scotese [11], slightly modified).

completely closed, forming the Ural Mountains and Laurasia. At the same time, the continent of South America and the southern part of Laurasia joined together, the Rheic Ocean closed and the continent of Gondwana joined the Proto-Tethys. The European and American

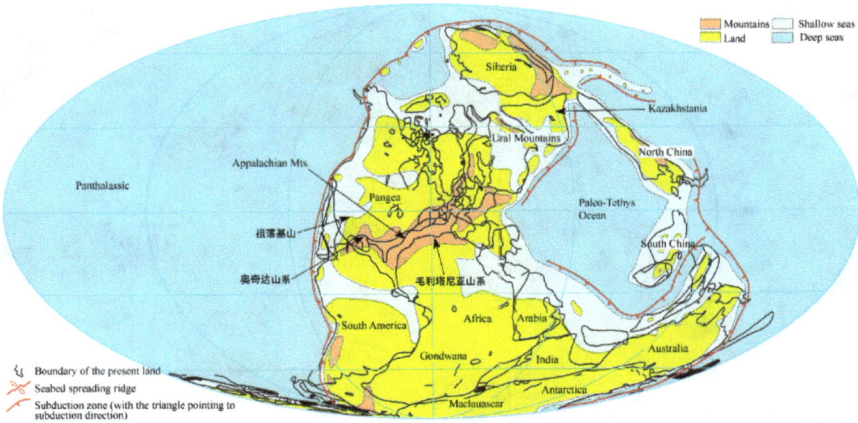

Figure 7.   Global plate remodeling map during the Late Carboniferous (306 Ma) (based on Scotese [11], slightly modified).

continents were connected to the Gondwana continent in the south, forming the western half of Pangaea. At this time, the southern pan-continent was covered by glaciers and swamp coal developed near the equator (Figure 7).

At the end of the Paleozoic, lands joined together to form Pangea, which was centered on the equator and extended from the Antarctic region to the North Pole. The Panthalassic Ocean and Paleo-Tethys Ocean were separated on the east and west sides. At this time, some continents (South China, North China, and Cimmeria) were still separated from the pan-continent.

In the Late Permian, the Cimmerian continent was separated from the Gondwana continent, then was connected with the Chinese continent, and later migrated toward the Laurian lands. At the same time, the Tethys Ocean began to form at its southern end, the Paleo-Tethys Ocean began to close, and the pan-continent reached its maximum extension range in the Late Permian (258 Ma) (Figure 8). At the end of the Permian, the earth suffered the largest biological extinction event in history and 99% of the creatures went extinct, marking the end of the Paleozoic era.

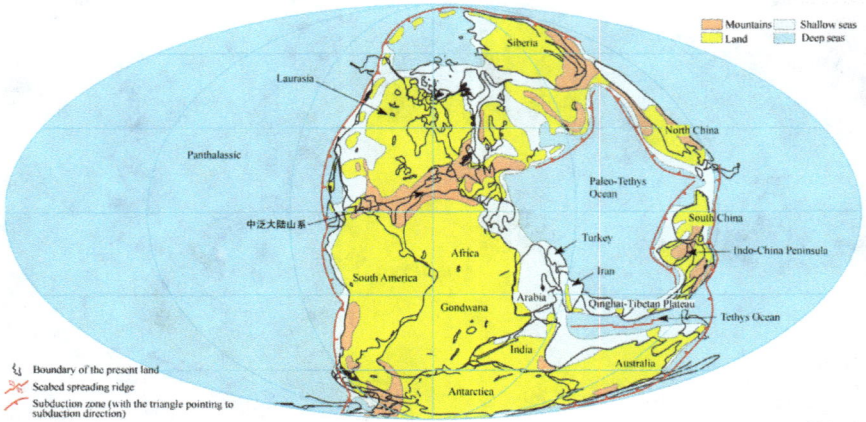

Figure 8.   Global plate remodeling map during the Late Permian (258 Ma) (according to Scotese [11], slightly modified).

2.4. *The Mesozoic*

During the Triassic, the pan-continent rotated to the southwest and the Cimmerian continent was accompanied by the shrinking Paleo-Tethys movement, which did not end until the Middle Jurassic. The Paleo-Tethys Ocean closed from west to east, resulting in the Cimmerian orogenic movement (Figure 9). In the Late Triassic, the Cimmerian continent and the southern edge of the Siberian continent joined together. At this time, all the landmasses merged into a true Pangaea [15]. The pan-continent was formed from the Devonian period and finally took shape in the Late Triassic. During the Triassic period, biological species began to evolve again after the extinction in the Permian. In the Early Jurassic (195 Ma), the pan-continent took a shape of the alphabet "C" and the Tethys Ocean was located in the "C" (Figure 10).

The breakup of the pan-continent can be divided into three stages. The first stage was the Middle Jurassic (175 Ma) when the pan-continent was split from the Tethys Ocean in the east and the Pacific Ocean in the west, finally forming Laurasia and Gondwana. Rifts occurred between the North American and African continents, accompanied by many disappearing rifts. The North Atlantic Ocean began to form in the rifts [16].

The Atlantic Ocean began to split from the north to the middle, and the southern part did not begin to form until the Cretaceous. The

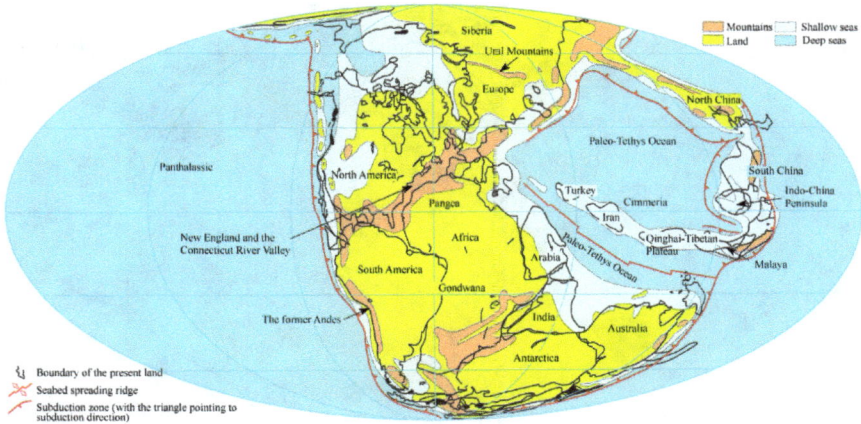

Figure 9. Global plate remodeling map during the Early Triassic (237 Ma) (according to Scotese [11], slightly modified).

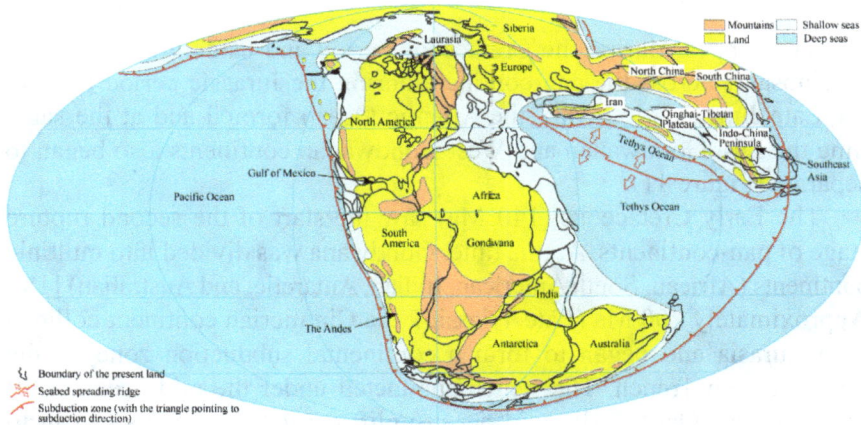

Figure 10. Global plate remodeling map during the Early Jurassic (195 Ma) (based on Scotese [11], slightly modified).

Laurasian continent rotated clockwise, with the North American continent moving northward and Eurasia moving southward, causing the Tethys Ocean to begin to close. At the same time, new rifts also occurred between the East African continent, the Antarctic continent, and the continent of Madagascar, causing the southwestern tip of the Indian

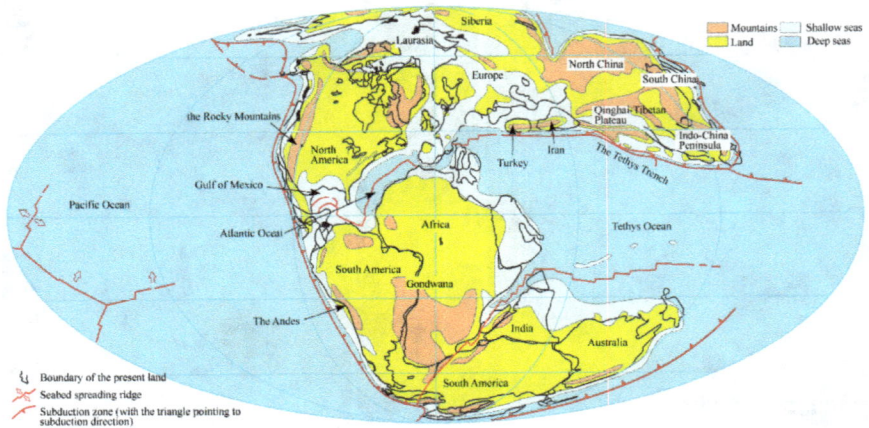

Figure 11.   Global plate remodeling map during the Late Jurassic (152 Ma) (according to Scotese [11], slightly modified).

Ocean to appear during the Cretaceous period and leading to the spread of dinosaurs across the pan-continent during the Jurassic period [17]. In the Late Jurassic (152 Ma), the Atlantic Ocean formed and at the same time the East Gondwanan and West Gondwanan continents also began to separate (Figure 11).

The Early Cretaceous (140 Ma) saw the start of the second rupture stage of pan-continents. At this time, Gondwana was divided into multiple continents (African, South American, Indian, Antarctic, and Australian) [18]. Approximately 200 Ma (Late Triassic), the Cimmerian continent collided with Eurasia and began to form a continental subduction zone — the Tethys Ocean Trench. The trench subducted under the mid-ocean uplift of the Tethys Ocean. The mid-ocean uplift caused the Tethys Ocean to expand, causing the African, Indian, and Australian continents to migrate northward. The expansion of the Atlantic Ocean in the Early Cretaceous made the Gondwanan continent disintegrate. The South American and African continents were finally separated from the East Gondwanan continent (Antarctica, India, and Australia), and the South Indian Ocean began to form. In the middle of the Cretaceous, with the South Atlantic rifting from south to north, the South American continent moved westward, away from the African continent. At the same time, the continents

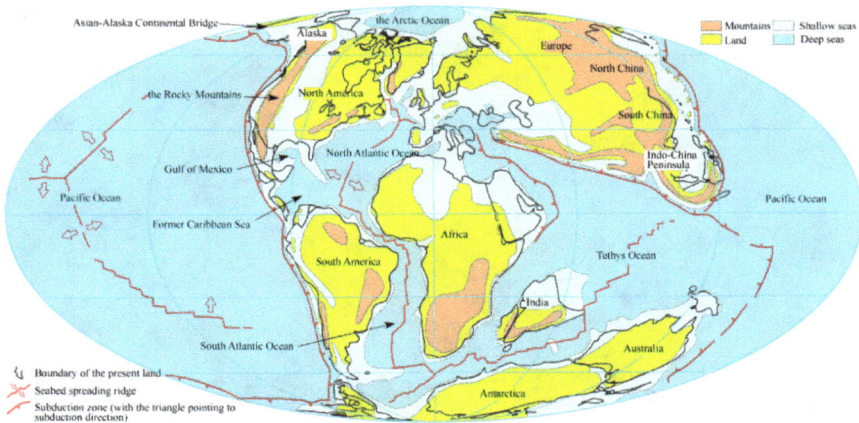

Figure 12. Global plate remodeling map during the Late Cretaceous (94 Ma) (according to Scotese [11], slightly modified).

of Madagascar and India began to separate from the Antarctic continent and migrated north, and the Indian Ocean formed. In the Late Cretaceous (100–90 Ma), the continents of Madagascar and India were separated from each other. The Indian continent continued to move northward, the Tethys Ocean closed, and the continent of Madagascar collided with the continent of Africa. At this time, the continent of North America was still connected to the continent of Europe, and the Australian continent still belonged to the Antarctic continent. The East Indian Ocean on the western edge of Australia had begun to rift and the ocean became wider. The Indian continent migrated to the southern edge of the Asian continent and was connected with the marginal island arc (Figure 12) [19].

During the Cretaceous, the global climate was warmer than today, similar to that of the Jurassic and Triassic. Meanwhile, shallow seas covered the most of the land, and the sea level was 100–200 m higher than today. This was mainly because of the rapid rifting of new oceans and the rapid expansion of mid-ocean ridges in the Cretaceous period, which caused sea levels to rise [20].

At some point from the end of the Cretaceous to the beginning of the Paleogene, the comet Chicxulub with a diameter of 16 km hit the earth, causing a sudden change in the earth's climate and the extinction of dinosaurs and other creatures (Figure 13).

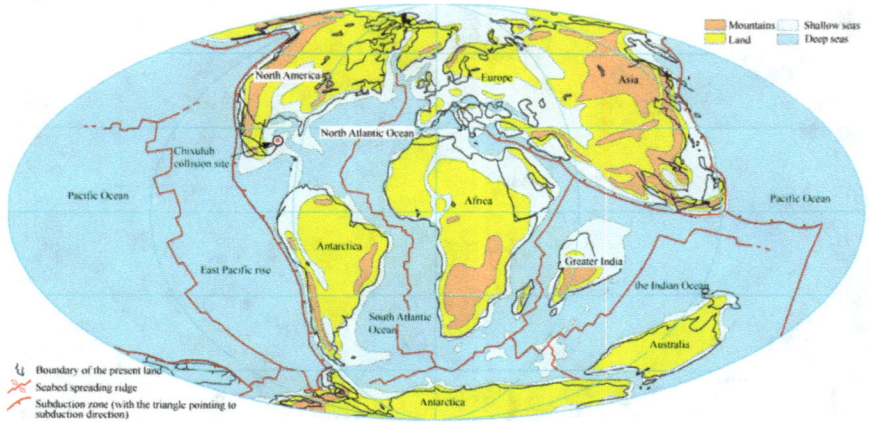

Figure 13.   Global plate remodeling map during the Cretaceous and early Tertiary (66 Ma) (according to Scotese [11], slightly modified).

2.5. *The Cenozoic*

The Early Cenozoic (the Paleocene-Oligocene) was the last stage of pan-continental depression. New ocean basins expanded and continents were squeezed violently to form depressions. During 60–55 Ma (the Paleocene-Eocene), Laurasia broke up, causing North America to break away from Eurasia. The Atlantic and Indian Oceans continued to expand and the Tethys Ocean continued to close. During 55–50 Ma (the Eocene), the Indian continent began to squeeze toward the Asian continent, leading to the formation of the Himalayas and the Qinghai–Xizang Plateau and also the closure of the Tethys Ocean (Figure 14). Today, these two land landmasses are still compressing and colliding with each other. Structures along the faults are active and seismic activity is still taking place. At the same time, the Australian continent was separated from the Antarctic continent and moved rapidly northward, now colliding with East Asia. The African continent migrated northwestward, approaching the European continent. [21, 22] During the Oligocene, the South American continent was separated from the Antarctic continent. At approximately 15 Ma (the Miocene), the main edge of the Australian continent collided with the southwestern part of the Pacific continent, prompting the formation of the New Guinea highlands (Figure 15).

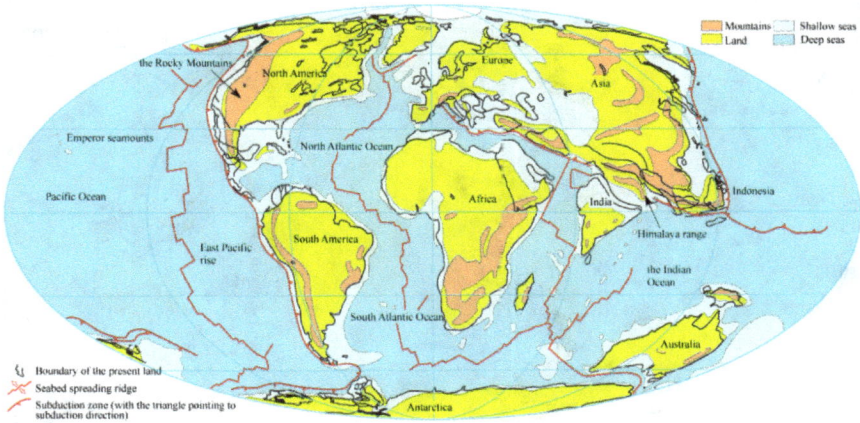

Figure 14. Global plate remodeling map during the Eocene (50 Ma) (according to Scotese [11], slightly modified).

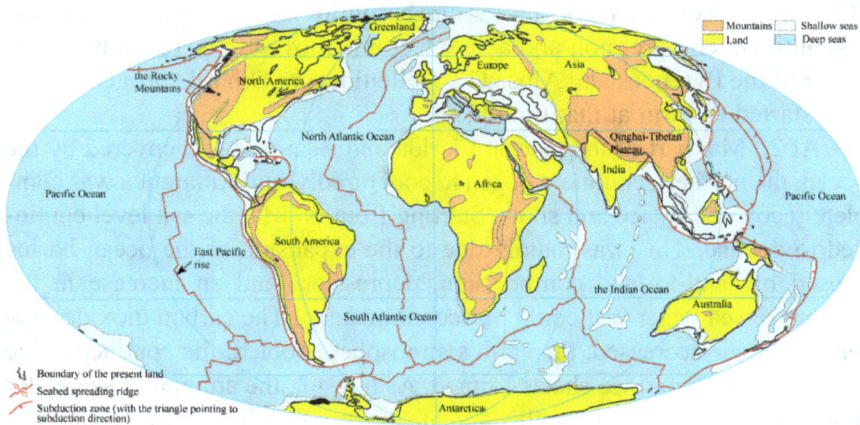

Figure 15. Global plate remodeling map during the Miocene (14 Ma) (according to Scotese [11], slightly modified).

During the Cenozoic, the pan-continental rift continued and many of the current rifting and disintegrating activities started at 20 Ma. The Red Sea began to form and the intrusion of sea water caused the Arabian Peninsula to separate from the African continent. The East African rift occurred and the Sea of Japan and the Gulf of California formed

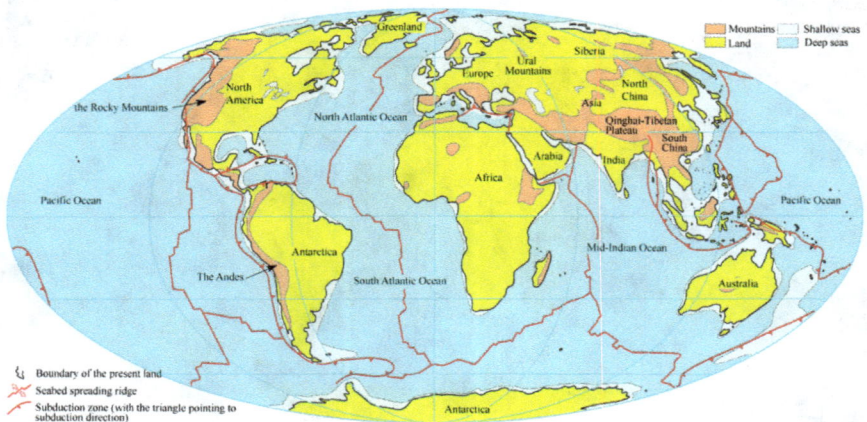

Figure 16.   Current location map of the global plate (according to Scotese [11], slightly modified).

(Figure 16). Active tension and fissure activities led to frequent continent-to-continent compression and active tectonic movements, resulting in the uplift of the Pyrenees and Alps. The Hellenide and Dinarde orogenic belts also started to form at that time [23].

At 20 Ma, part of the land in Florida and Asia was still covered by the ocean, the northern continent began to cool rapidly, and Antarctica was completely covered by ice and snow. Starting from 66 Ma, the sea level continued to decline. This was mainly due to the expansion of the ocean basins caused by land-to-land compression, depression, and an increase in the amount of sea water that could be accommodated. The carbon dioxide content in the air decreased, the earth's atmosphere cooled, the continental ice layer expanded, and ice sheets formed. As a result, the amount of sea water reduced by 5 Ma and the earth's climate entered the Great Ice Age again.

References

[1]   Compilation Group of Qinghai Province Stratigraphic Table. *Qinghai Book of Regional Stratigraphic Table of Northwest China*. Beijing: Geological Publishing House, 1980.

[2]   Qinghai Provincial Bureau of Geology and Mineral Resources. *Geology of Qinghai Province*. Beijing: Geological Publishing House, 1991.

[3] Dongzhou Qiu. *Triassic Lithofacies Palaeogeography in Asia and Central Pacific.* Beijing: Geological Publishing House, 1991.

[4] Yuanxi Qiu. *Tectonic Evolution of Yunkai Mountain and its Adjacent Regions.* Beijing: Geological Publishing House, 1993.

[5] Yuanxi Qiu. *Tectonic Properties of Snow Uplifts and its Superimposed Transformation on Paleozoic Oil and Gas Basins on Southeastern Margin of Upper Yangtze River.* Beijing: Geological Publishing House, 1993.

[6] Yuanxi Qiu. *Basic Characteristics and Formation Mechanism of Mesozoic and Cenozoic Composite Deformation Metamorphic Zones in the Lianhua Mountain Fault Zone in Guangdong Province.* Beijing: Geological Publishing House, 1991.

[7] Jishun Ren. *Tectonics and Evolution of China's Land.* Beijing: Science Press, 1980.

[8] Shanxi Provincial Bureau of Geology and Mineral Resources. *Regional Geology of Shanxi Province.* Beijing: Geological Publishing House, 1989.

[9] Shaanxi Provincial Bureau of Geology and Mineral Resources. *Regional Geology of Shaanxi Province.* Beijing: Geological Publishing House, 1989.

[10] Dianqing Sun. *Geomechanics Theories and Practices in Petroleum Survey and Exploration in China.* Beijing: Geological Publishing House, 1989.

[11] Scotese, C. R. Paleogeographic Atlas, PALEOMAP Progress Report No. 90-0497, 1997, University of Texas at Arlington, Arlington.

[12] Hongzhen Wang. *Tectonic Palaeogeography and Bio-palaeogeography of China's Neighboring Regions.* Wuhan: China University of Geosciences Press, 1990.

[13] Jun Wang, Dongpo Wang, Ushakov, *et al. Formation and Evolution of Sedimentary Basins in Northeast Asia and their Petroleum Prospects.* Beijing: Geological Publishing House, 1997.

[14] Zhixin Wang, Zhijun Jin. *Petroleum Geological Characteristics of Siberian Platform and its Marginal Depressions.* Beijing: China Petrochemical Press, 2007.

[15] Guoping Bai. *Petroleum Geological Characteristics of Oil and Gas Fields in Middle East.* Beijing: China Petrochemical Press, 2007.

[16] Tingyu Chen, Yanbin Shen, Yue Zhao, *et al. Geological Development of Antarctica and Evolution of Gondwana Ancient Land.* Beijing: Commercial Press, 2008.

[17] Wenxin Zhong, Yundong Zhang. *World Oil-producing Countries (South America).* Beijing: China National Petroleum Corporation, 1998.

[18] Xuexiang, Y., Dianyou, C. Tectonic movement and global climate change. *Journal of Geoscientific Research in Northeast Asia,* 2000 3(2): 121–128.

[19]   Xue-Feng, Z., Yan-De, Z., Ming-Ji, Z. Differential tectonic movement of Yanchang Formation in southwestern margin of Ordos Basin and its geologic significance. *Lithologic Reservoirs*, 2010 22(3): 78–79.

[20]   Yue-Zhong, W. U. Tectonic attribution and movement characteristic of Altyn mountain. *Journal of Earth Sciences and Environment*, 2008 2(1): 111–117.

[21]   Xiaoguang Tong, Zengmiao Guan. *Atlas of World Petroleum Exploration and Development*. Beijing: Petroleum Industry Press, 2004.

[22]   Xiaoguang Tong. *Atlas of World Petroleum Exploration and Development (Commonwealth of Independent States)*. Beijing: Petroleum Industry Press, 2004.

[23]   Jiashu Wang. *Timan-Pechaola Petroliferous Basin*. Beijing: Petroleum Industry Press, 1991.

Chapter 8

Conclusions

Under the action of geodynamics and various *in-situ* stresses, the global crust has produced eight tectonic styles in different geological historical periods, but the overall style is uplift and subsidence depression. Lands have appeared as a result of uplifting of the earth, whereas oceans have formed as a result of sinking. The uplifts on the continents turned into mountains, whereas the depressions turned into lakes or swamps. The crust has evolved in this way since beginning of the earth's existence. There has been no continental drift or plate movement. Crustal fault activities, volcanic eruptions, and seismic activities were the manifestations of the imbalanced effects of local ground stress, but they were local phenomena, which could not cause the separation of the crust at all. Thus, the earth's crust will always be a complete and indivisible whole as it has always been in the past.

Therefore, the global crust can only be a product of the evolution of uplifts and depressions as well as sea–land changes.

Extended Reading

1. Introduction

This book's main body will enrich and further the development of global geological science based on the original geomechanics theory. To enhance readers' understanding of the comprehensive and in-depth theoretical contents of this book, we, the author's team members, decided to include an Extended Reading section. This section comprises five relevant articles, originally published in Chinese and now translated into English for the first time to complement the main text.

We begin this section with the article "Evolution of Global Crustal Uplift, Subsidence, and Basins," which will help broaden and deepen readers' understanding of hydrocarbon distribution and the structural deformation style stated in previous chapters.

The four articles that follow — "Features of Structural Systems in Northern China and Their Control on Basin and Hydrocarbon Distribution," "A Study on an Oil Control Model of a Subordinate Shear Structural System in China," "Tectonic Systems in Northwestern China and Their Relations with Hydrocarbon," and "Petroleum Control Patterns using Structural Systems" — reveal the breakthroughs and discoveries made under the guidance of the conclusions of the previous chapters, and prove the scientific reasoning and accuracy of the theories applied to guide China's oil and gas exploration works effectively.

We hope this section opens readers' minds about the importance of the global crustal uplift and depression movements and applying theory into practice.

Evolution of Global Crustal Uplift, Subsidence, and Basins

Abstract

The earth's crust is under various stresses, including compressive stress, tensile stress, and torsional stress, resulting in uplift, subsidence, depression, and strike-slip. Structural movement causes uplifts and depressions in the crust, leading to changes in the land and sea. During the geological evolutionary history of the earth, the crust uplifted into orogenic zones or uplift areas, while depressions led to various types of basins. Basins across the world can be divided into five major types, namely, rift-craton, intracratonic depression, foreland, fault, and depression basins. There are eight major deformation styles in the global structure, namely, E-W-trending, N-S-trending, N-E-trending, N-N-E-trending, N-W-trending, epsilon-shaped, S- or reverse-S-shaped, and rotation-torsional deformation. In Paleozoic cratonic basins of China, hydrocarbon is mainly distributed in paleo-uplifts, paleo-slopes, regional unconformities, and fault zones; in Mesozoic and Cenozoic faulted basins, hydrocarbon is mainly distributed in steep slopes, gentle slopes, and central structural zones; and in Mesozoic and Cenozoic foreland basins, hydrocarbon is distributed in fault fold zones, slope zones, and overthrust zones. Various twisted structures, such as broom-shaped, echelon, rotation-torsional, reverse-S-shaped, and λ-shaped structures, control hydrocarbon distribution.

Keywords: Hydrocarbon distribution, structural deformation style, uplift and depression evolution, basin type, global structure.

117

For many years, the law of the change in the earth's crustal structure has been discussed on the basis of the features of global crustal structural deformation [1, 2]. Due to the self-rotational force of the earth, the impact of celestial bodies on the earth, the presence of radioactive substances on the earth, the thickness and density of various parts of the earth's crust, and other factors, various crustal stresses of different scales, directions, and properties have appeared. These complex crustal stresses have caused structural deformation of the earth's crust, mainly showing evolution in uplift and depression. The crustal uplifts have become orogenic zones or uplift areas, while the crustal depressions have become various types of basins. Basins across the world can be divided into five major types, namely, rift-craton, intracratonic depression, foreland, fault, and depression basins. There are eight major deformation styles in the global structure, and different basin types and structural styles control hydrocarbon distribution.

1. Major Global Structural Movements

The structural evolution of the crust and lithosphere is driven by the main structural movements of the crust, and the formation, evolution, and distribution of the structure is influenced by the crustal movement mode, the period of time of the movement, and its change law.

   The main structural movements in China and the world are shown in Table 1.

2. Main Forms of Crustal Deformation

There are three main forms of crustal deformation, namely, depression (basin) formed by a stretching movement, uplift formed by a compression movement, and depression and uplift (basin) formed by a strike-slip movement. Uplift and depression alternate with each other. Strong compression occurs in an uplift area, which will inevitably lead to large-scale depression in the adjacent areas. Through theoretical analysis and practical investigations, it has been recognized that folded orogenic zones have been subjected to strong horizontal compression, which can be provided by large-scale subsidence. Because of the large-scale subsidence process, under the transmission of force, heat energy, gas, liquid, and magma in deeper positions are forced to move horizontally in a parabola form with

Table 1.   Main structural movements in the world.

Geological age		Isotope/age value/Ma	Major geological events	Tectonic stage and crustal movements		
				Europe and America	China	Africa
Cenozoic	Quaternary period — Holocene	Present 0.01		New Alpine stage	Himalayan stage	Alpine stage
	Quaternary period — Pleistocene	2			Himalayan Movement (Late)	Late Alpine Movement
	Neogene period — Pliocene epoch	5	Disintegration stage of the united continent	Saff Movement ~~~		
	Neogene period — Miocene epoch	22.5		Pyrenean Movement ~~~	Himalayan Movement (Early)	Early Alpine Movement
	Paleogene — Oligocene	37.5		Laramide Movement ~~~	Yanshan Movement (Late)	
	Paleogene — The eocene epoch	50		The New Simerian Movement	Yanshan Movement (Middle)	
	Paleogene — Paleocene	65			Yanshan Movement (Early)	Hercynian Stage
Mesozoic	The cretaceous period	137		Alpine stage	Indosinian movement (early)	Third act of Hercynian Movement
	Jurassic Period	185		The Ancient Simerian Movement	Indosinian movement (early)	Second act of Hercynian Movement
	Triassic period	230			Yining movement	First act of Hercynian Movement
Neopaleozoic	Permian	280		Hereynian stage		
	Carboniferous	350	Formation stage of united ancient land	Appalachian Movement	Tianshan Movement	
	Devonian	400		Breton Movement ~~~		Pan–African phase
Early paleozoic era	Silurian period	440		Caledonian Movement	Qilian (Guangxi) Movement	(Katanga) Late Pan-African Movement
	Ordovician period	500		Erian Movement ~~~	Caledonian Movement — Gulang Orogeny	
	Cambrian period	610		Taikang Movement ~~~	Xingkai Orogeny	
Proterozoic	Sinian period	850		Anita Movement ~~~	Jinning Movement (Late)	(Katanga) Early Pan-African Movement
	Neo-...	1055	Stage of platform formation	Goethe and Greenville Movement ~~~	Luliang Jining stage — Jinning Movement (Early)	
	Meso-...	1600–1700		Cary Hudson Movement ~~~	Luliang (Middle) Movement	
	Paleo-...	2500–2600			Fuping Changli stage — Wutai Orogeny	
	New	2900–3000		Sam-Kennel Movement ~~~	Fuping Movement	
Archaean	Ancient	3800	Stage of continental core formation			
Prearchaean		4600	Astronomical phase			

the top of the arc downward, and invade and rise to the adjacent areas, forming a substantial uplift of convection and collision under the opposite direction in the deeper position of uplift zone, which is a basic and important feature of crustal movement dominated by subsidence.

Lateral pressure can be produced in the settlement process, easily resulting in strong deformation, fold, and thrust. Geosyncline evolution occurs in the early stage of the return phase, which can be illustrated by various phenomena. The geophysical data show that the edges of oceans such as the Atlantic Ocean, the Indian Ocean, and the Arctic Ocean were unconformably superimposed on the structures of the adjacent continents before the Mesozoic, fully indicating the result of strong horizontal dynamic movement caused by the large-area subsidence process of ocean basins.

The common compound anticlines and synclines in regional geology appear repeatedly and alternately, among which the strata with compound anticlines are strongly folded and uplifted, forming igneous rocks and fractures, while synclines become the subsidence depressions, leading to basins.

3. Eight Structural Deformation Styles in the World

There are eight structural deformation styles [1–5] as follows: (1) E-W-trending one; (2) N-S-trending one; (3) N-E-trending one; (4) N-E-trending one; (5) N-W-trending one; (6) Epsilon-shaped one; (7) S-shaped or reverse-S-shaped one; and (8) Rotation-torsional one. Among them, the E-W-trending and N-S-trending styles dominate.

The evolution of global structural deformation styles has five features, namely, stage, inheritance, difference, migration, and transformation, showing the complexity of the various structural deformation styles. In summary, uplift and depression are the main features of crustal deformation.

The structural deformation styles control the formation and evolution of large and small landmasses in all continents, while these landmasses control and affect the formation and evolution of structural deformation styles, and their interactions have created the current global structural framework and land–sea changes.

4. Basin Types and Hydrocarbon Distribution Law

After years of research, global basins have been divided into five types, namely, rift-craton, intracratonic depression, fault, foreland, and depression basins. Taking the basin types developed in China as representatives, the hydrocarbon distribution law is discussed here [6–11] (Table 2).

4.1. *Hydrocarbon Distribution in the Paleozoic Craton Basins*

The Chinese craton basins are mainly distributed on the China–North Korea Platform, the Yangtze Platform, the Tarim Platform, and the

Table 2.   Evolution stages of the Sinian–Paleozoic basins in China.

Period	Evolution stage	Evolution characteristics
Sinian–Middle Ordovician	Rifting and oceanic expansion (rift-craton Basin)	The North China, South China, Tarim, Junggar, and Qinghai–Xizang landmasses were separated and surrounded by rifted basins and craton basins (North China and South China biota)
Late Ordovician–Middle Silurian	Compressional subduction of land mass (flexural craton)	Due to the north–south compression, the ocean basins gradually diminishing in the continental blocks. The North China block became a paleo-land (O_3-C_1 missing) and mainly formed compressive craton basins
Late Silurian–Devonian	Compression collision orogeny and formation of the paleo-continent of China (flexural craton)	Uplift orogeny formed at the edges of continental blocks, the Foredeep basins formed at the front of orogenic zones, and the ancient Asian Ocean expanded continuously
Carboniferous–Permian	Tension-extrusion uplift (intracratonic depression basin)	In the C-P_1 Period, the Paleotethys Ocean spread (in the southwest), the sea level rose, and the sea and land interdeposited in China widely; there were intense volcanic activities in the P_1 Period and compression in the P_2 Period; the seawater in Tarim, Junggar, Qaidam, and North China withdrew and became continental sediment; the South China Sea facies lasted until the end of the T_2 Period, forming extensive intracraton depression basins

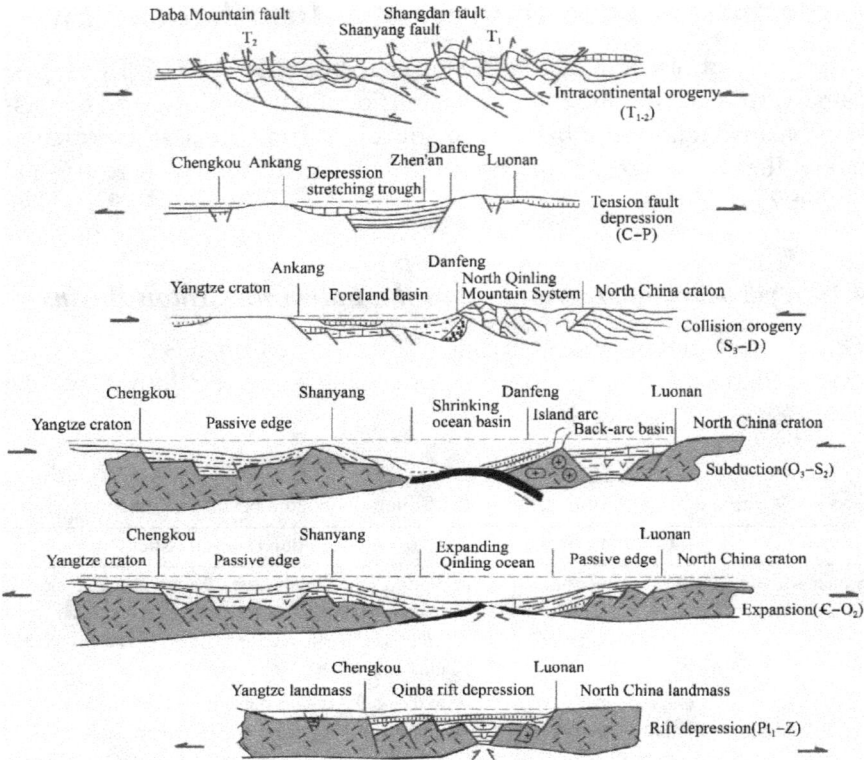

Figure 1.   Evolution of the Paleozoic cratonic basins in China (see Ref. [2]).

Junggar Platform (Figure 1) [12], and hydrocarbon is mainly distributed in the paleo-uplifts, paleo-slopes, regional unconformities, and fault zones of the related basins (Figure 2).

4.1.1. *Paleo-uplift*

A paleo-uplift refers to an uplift area that was deposited in the Paleozoic, which is also in the direction of hydrocarbon migration, such as the Shaya Uplift and the Kartak Uplift in the Tarim Basin (Figure 2); the Luliang Uplift in the Junggar Basin; the Wushenqi Uplift and the Eastern Uplift in the Ordos Basin; and the Leshan Uplift, the Kaijiang Uplift, the Jiangyou Uplift, and the Daxing Uplift in the Sichuan Basin (Figure 3).

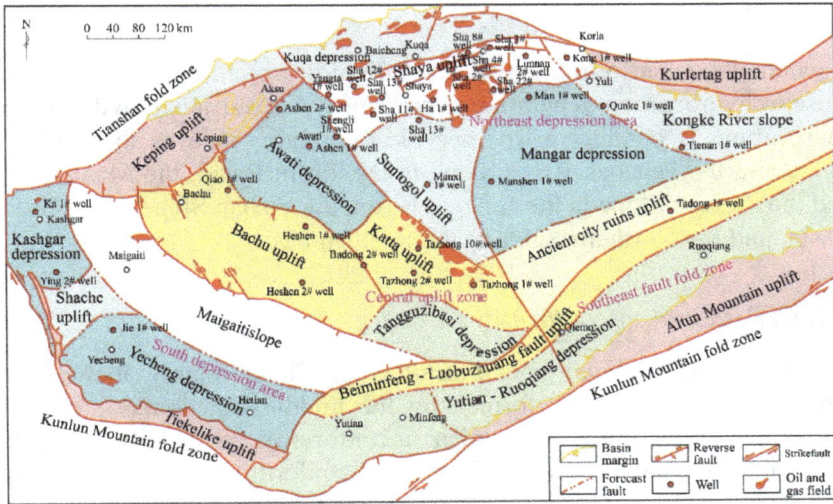

Figure 2. Hydrocarbon distribution in the Tarim Basin.

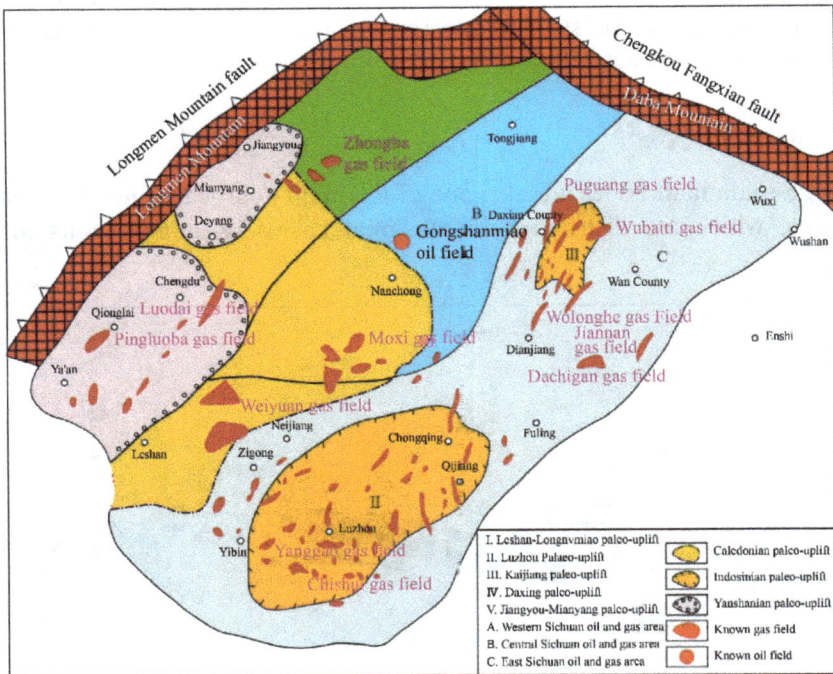

Figure 3. Hydrocarbon distribution in the Sichuan Basin.

4.1.2. *Paleo-slope*

A paleo-slope is also in the direction of hydrocarbon migration, which is close to hydrocarbon-generating depressions with sufficient hydrocarbon sources and is conducive to reservoir formation, such as the Maigaiti Slope and the Peacock River Slope in the Tarim Basin, the Mahu Slope and the Zhunan Slope in the Junggar Basin, the Yishan Slope in the Ordos Basin, and the Northern Slope in the Sichuan basin.

4.1.3. *Regional unconformity*

The Paleozoic regional unconformities in China mainly include the Ordovician unconformity formed by the middle Caledonian structural Movement, the Silurian–Devonian unconformity formed by the early Hercynian Movement, and the Permian unconformity formed by the late Hercynian Movement. The sandstone conglomerate above the unconformable surfaces and the carbonate weathering crust below the unconformable surfaces all have good reservoir properties and can be used as good channels for hydrocarbon migration, such as the Awati Depression and the Mangar Depression in the Tarim Basin, which are controlled by unconformity surfaces (Figure 4).

4.1.4. *Fault zone*

Large-scale fault zones are also one of the main hydrocarbon enrichment zones, which are distributed in the great basins. Regional faults that

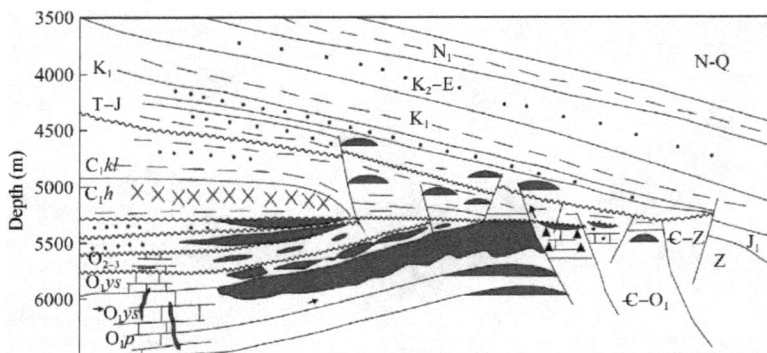

Figure 4.   Oil controlling mode of fault unconformities in the Tahe Oil Field.

developed on the edge of the paleo-uplifts generally have features such as long active histories and large-scale and wide fault fractures, which mainly play a role in the vertical migration of hydrocarbon and constitute regional hydrocarbon source faults.

4.2. *Hydrocarbon Distribution in the Mesozoic and Cenozoic Faulted Basins*

The Mesozoic and Cenozoic faulted basins are mainly distributed in eastern China [3, 13–20] and were formed by collisions with the Indian Plate and large-scale repulsive movements. During this period, under the expansion of the marginal seas in the east and the movement of material creep in the deep positions of the Chinese mainland, the past state changed and the whole of the mainland extended and expanded toward the seas; downward pressure and upward tension were formed on the east of the mainland based on the uplifts and an intracontinental faulted basin was formed in the crustal thinning process.

Hydrocarbon is mainly distributed in steep slope zones, gentle slope zones, and central structural zones (Figure 5). A steep slope zone is the starting zone of extension activities in the faulted basin and is also the development site of the main fault for depression control. Steep slopes, close provenance, large undulations in paleo topography, narrow phase zones, rapid changes, and intense tectonic activities characterize the steep-slope zone. A gentle slope zone is an important part of dustpan depression

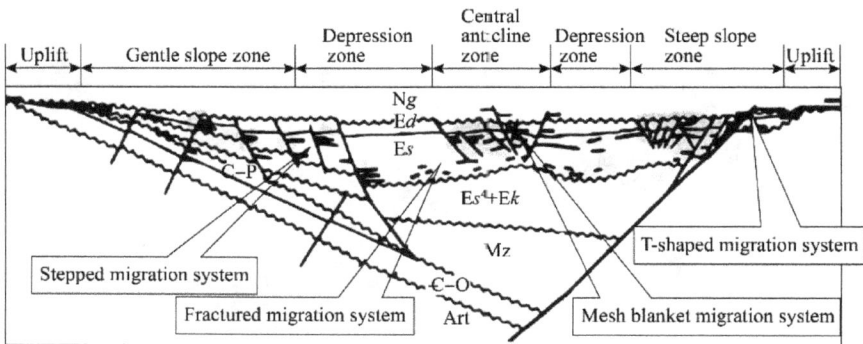

Figure 5.   Structural section and hydrocarbon distribution in the Bohai Bay Basin.

in the faulted basin and an important direction for hydrocarbon migration. A gentle slope zone is simple in structure, and generally has a nose. If it is blocked in the upward direction, an oil reservoir can be formed. Generally, there are two groups of faults, and a bedrock block is cut into several secondary tilting fault blocks, which are beneficial to the formation of buried hills and hydrocarbon reservoirs at the fault edge of the tilting fault blocks with the formation of stratigraphic overlapping hydrocarbon reservoirs at the waist of the fault blocks. The central anticline zone is formed by plastic arching of strata caused by fault activities. Many groups of faults are often developed at the top of the anticline and have very complex structural features, which can form many types of traps (anticline, nose structure, fault block, etc.).

Generally, faulted basins have sedimentary features, such as turbidite fans, alluvial fans, delta fans, and submarine fans, which are beneficial to form lithologic or structural–lithologic traps combined with a distribution of faults.

4.3. *Hydrocarbon Distribution in the Mesozoic and Cenozoic Foreland Basins*

The Miocene and Cenozoic foreland basins are mainly distributed in the central and western areas [11, 21–24], and hydrocarbon in foreland basins is mainly distributed in three structural zones: fault fold zones, slope zones, and overthrust zones (Figure 6).

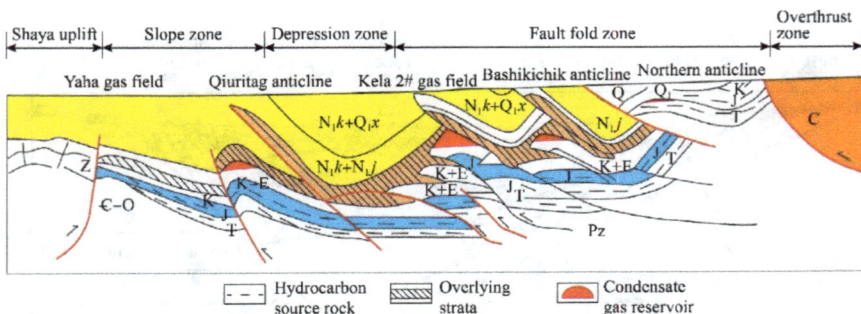

Figure 6.   Structural section and hydrocarbon distribution in the Kuqa Foreland Basin of the Tarim Basin.

4.3.1. *Foreland fault fold zone*

Hydrocarbon is mainly distributed in row 2 and row 3 of the fault fold zone, and oil is mainly contained in the anticline, especially in the footwall anticline controlled by the overthrust fault between the dual structures, such as the Hutubi Gas Field in the Junggar Basin, the Kela 2# Gas Field in the Tarim Basin, and the Lenghu Oil Field in the Qaidam Basin.

4.3.2. *Foreland slope zone*

The slope zone is located in the direction of hydrocarbon migration with relatively stable structural facies and is also a favorable zone for hydrocarbon enrichment, such as the Yaha Hydrocarbon Field in the Tarim Basin and the Yongjin Hydrocarbon Field in the Junggar Basin.

4.3.3. *Foreland overthrust zone*

The geological conditions in this zone are complex. With the development of 3D seismic and drilling technology, the structural features and trap conditions underlying thrust faults can be identified, due to which more hydrocarbon fields are gradually being discovered, such as the Qingxi Oil Field in the Jiuxi Basin and the Shenghe 1# Well in the Tarim Basin.

4.4. *Hydrocarbon Controlled by All Kinds of Twisting Structures*

Under the guidance of the theory of geomechanics, through studies of several basin structural systems [17, 25–32], it has been recognized that hydrocarbon distribution fields are mainly controlled by various low-order torsional structural systems, with typical features such as broom-shaped, echelon, rotation-torsional, reverse-S-shaped, and λ-shaped structures.

4.4.1. *Broom-shaped structural zone*

The Tazhong broom-shaped structural system is located on the Kartak Uplift in the middle of the central uplift zone of the Tarim Basin, and is a broom-shaped structure that spreads northward and converges to the southeast (Figure 7). At present, industrial hydrocarbon discoveries in the

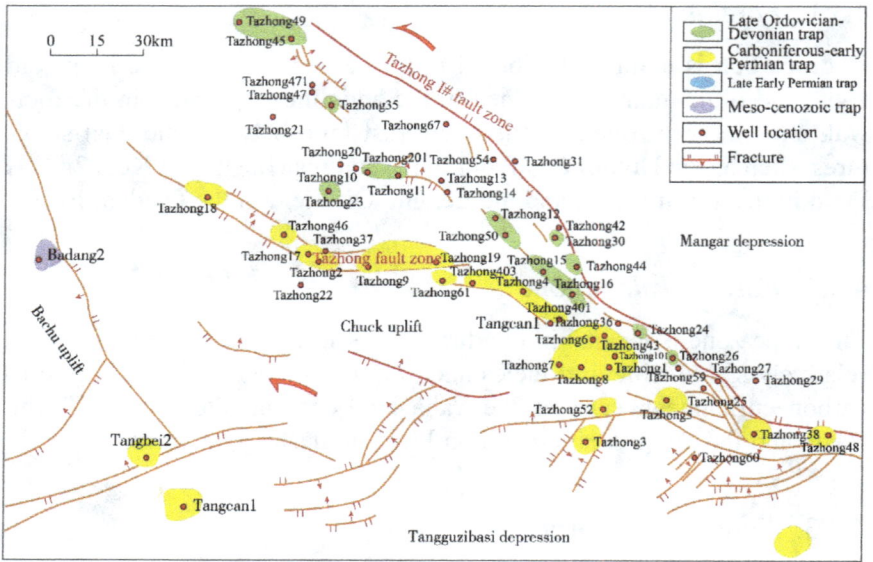

Figure 7.   Broom-shaped structure section and hydrocarbon distribution in the Tazhong area of the Tarim Basin.

Tazhong hydrocarbon area are mainly concentrated in the eastern part of the Tazhong 1# and Tazhong 2# fault zones and the northern slope of Tazhong.

4.4.2. *Echelon structural zone*

A typical example of the echelon structural zone is the Yecheng Echelon Structural System in the southwest depression of the Tarim Basin. The system is located in the Yecheng Depression formed in the Himalayan Period; it was formed by the compression and torsion of the Kunlun Orogenic Zone in the basin and consists of three rows of structures.

4.4.3. *Rotation-torsional structural zone*

A typical rotation-torsional kind of structure is located on the Akkule Uplift in the middle section of the Shaya Uplift in northern part of the

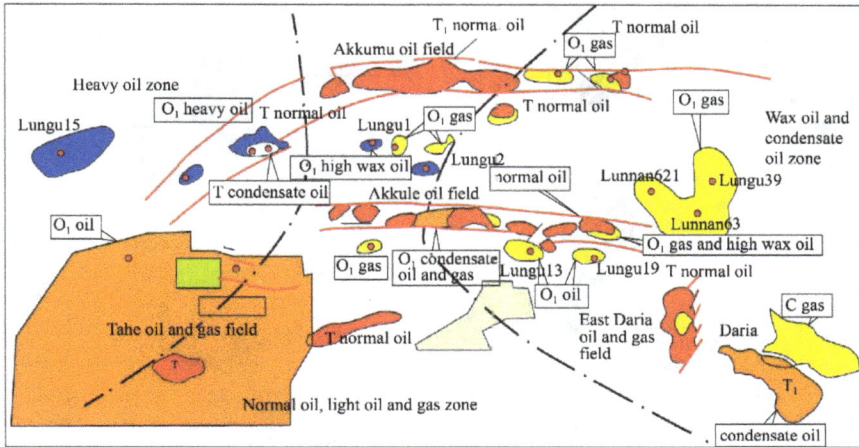

Figure 8. Rotation-torsional structure section and hydrocarbon distribution in the Tahe area of the Tarim Basin.

Tarim Basin, formed in the Hercynian Period, and consists of two rotation-torsional structural zones and eddies; this zone plays an important role in controlling the hydrocarbon distribution fields in this area (Figure 8).

4.4.4. *Reverse-S-shaped structural zone*

The reverse-S-shaped structural system of Lenghu is located in the structural uplift between the Saishiterg Depression and the Kunteyi Depression in the Saikun Fault Depression in the northern margin of the Qaidam Basin, and consists of the anticlines of Lenghu 0~7#, Nanbaxian, Beilingqiu, Donglingqiu, and Nanjixing of various scales. The hydrocarbon fields are mainly distributed at the turning points of the reverse-S-shaped structural system, which are the torsional structural parts with moderate *in-situ* geostress (Figure 9).

4.4.5. *λ-shaped structural zone*

The Bashituo λ-shaped structural system is located in the northern part of the Maigaiti Slope in the Tarim Basin. It was formed in the

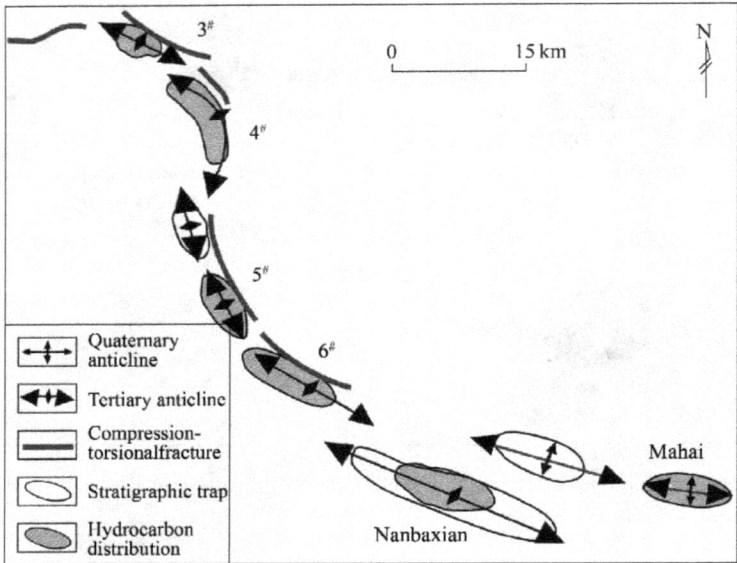

Figure 9.   Reverse-S-shaped structure and hydrocarbon distribution in the Qaidam Basin (see Ref. [33]).

N-W-trending Selbuya Fault Zone and the Bashituo Structural Zone in the late Hercynian Period and has been confirmed as a torsional structure. The Bashituo Hydrocarbon Field was formed in the Himalayan Period, and hydrocarbon came from the Carboniferous–Permian System, mixed with a large number of Paleozoic Cambrian–Ordovician components, which is typical for λ-shaped structural systems.

4.4.6. *Imbricate-shaped fault zone*

Imbricate-shaped fault zones are composed of several reverse faults that occur close together. The upper wall of each fault is distributed in a rising state, which is similar to the overlapping of roof tiles. Hydrocarbon reservoirs in the imbricate-shaped fault zone are roof-shaped vertically and imbricate-shaped horizontally; oil-bearing sections are densely concentrated, with many layers, and the reservoirs are distributed in narrow strips along the fault direction.

5. Conclusion

During the geological evolutionary history of the earth, the earth's crust was subjected to various geostresses, resulting in crustal deformation, showing uplift and depression. Uplift and depression are the main bases for the changes between the ocean and the land; that is to say, the land became subsided and depressed to form the ocean, and the crust under the ocean rose to form the land, but the overall evolution involved uplift and depression. Crustal uplift becomes an orogenic zone or uplift area, when the depression becomes basins of various types. The basins across the world can be divided into five major types, namely, rift-craton, intracratonic depression, foreland, fault, and depression basins. There are eight major deformation styles in the global structure, namely, E-W-trending, N-S-trending, N-E-trending, N-N-E-trending, N-W-trending, epsilon-shaped, S- or reverse-S-shaped, and rotation-torsional deformation. The different basin types and structural styles control the distribution of hydrocarbon.

References

[1] Yuzhu Kang, Shuwen Xing. *Introduction to Global Tectonic System*. Beijing: Geological Publishing House, 2018.
[2] Yuzhu Kang. *Global Hydrocarbon Distribution and Development Strategy*. Beijing: Geological Publishing House, 2016.
[3] Yuzhu Kang, Zongxiu Wang, Huijun Li, *et al*. *Study on Oil Control of Structural System in Songliao Area*. Beijing: Geological Publishing House, 2015.
[4] Yuzhu Kang, Zongxiu Wang. *Oil Control of Structural System in North China Basin*. Beijing: Geological Publishing House, 2017.
[5] Yuzhu Kang, Zongxiu Wang. *Study on Oil Control of Structural System in Sichuan Basin*. Beijing: Geological Publishing House, 2014.
[6] Baojun Liu, Xiaosong Xu, Xing'nan Pan, *et al*. *Sedimentary Crust Evolution and Mineralization of the Paleo-continent in Southern China*. Beijing: Science Press, 1993.
[7] Yuzhu Kang. Characteristics of petroleum geology setting and hydrocarbon prospects of Northwestern China. *Petroleum Geology & Experiment*, 1984 6(3): 229–240.
[8] Yuzhu Kang, Xiyuan Cai, Chuanlin Zhang. *Formation Conditions and Distribution of Paleozoic Marine Hydrocarbon Fields in China*. Urumqi: Xinjiang Science and Technology Health Press, 2002.

[9]   Yuzhu Kang, Hongjun Sun, Zhihong Kang, *et al. Paleozoic Marine Petroleum Geology of China*. Beijing: Geological Publishing House, 2011.

[10] Yuzhu Kang. Characteristics of Paleozoic marine oil generation in China. In *Scientific and Technological Progress and Social and Economic Development Facing the 21st Century*, Volume I. Urumqi: Xinjiang Science and Technology Health Press, 1999.

[11] Yuzhu Kang, Zongxiu Wang, Zhihong Kang, *et al. Study on Oil Control of Structural System in Junggar — Turpan Basins*. Beijing: Geological Publishing House, 2011.

[12] Yuchang Zhang. *Prototype Analysis of Petroliferous Basins in China*. Nanjing: Nanjing University Press, 1997, p. 48.

[13] Ci Bao, Xianjie Yang, Dengxiang Li. Geological structure characteristics and gas prospect prediction of the Sichuan Basin. *Natural Gas Industry*, 1985 5(4): 1–11.

[14] Jiqing Huang, Jishun Ren, Chunfa Jiang, *et al. Geotectomics and its Evolution in China*. Beijing: Science Press, 1980.

[15] Guangming Zhai. *Petroleum Geology of China (Vol. 13): Yumen Oil Field*. Beijing: Petroleum Industry Press, 1989.

[16] Guangming Zhai. *Petroleum Geology of China (Vol.14): Qinghai–Xizang Hydrocarbon Region*. Beijing: Petroleum Industry Press, 1990.

[17] Yuzhu Kang. *Petroleum Geomechanics*. Beijing: Geological Publishing House, 2012.

[18] Dianqing Sun. Geomechanics theory and practice in petroleum exploration in China. *Bulletin of the Institute of Geomechanics CAGS*, 1989 1(13): 9–22.

[19] Xi Xu, Xiaoying Zhu, Xipeng Shan, *et al.* Structure and sedimentary characteristics of the Meso-Cenozoic Basin group along the Yangtze River in the Lower Yangtze region. *Petroleum Geology & Experiment*, 2018 40(3): 303–314.

[20] Weizhong Zhang, Yunyin Zhang, Xingmou Wang, *et al.* Transmission model of secondary gas reservoir on the Basin margin of Jiyang Depression. *Petroleum Geology & Experiment*, 2019 41(2): 185–192.

[21] Yuzhu Kang, Zongxiu Wang. *Oil Control of Structural System in Tarim Basin*. Beijing: Geological Publishing House, 2009.

[22] Yuzhu Kang, Zongxiu Wang. *Oil Control of Structural System in the Qaidam Basin*. Beijing: Geological Publishing House, 2011.

[23] Yuzhu Kang, Zongxiu Wang. *Study on Oil Control of Structural System in the Ordos Basin*. Beijing: Geological Publishing House, 2014.

[24] Zonghao Li, Hailei Liu, Baoli Bian, *et al.* Structure characterization and exploration potential analysis of the Shielding zone in the Northwestern Rim of Junggar Basin. *Special Oil & Gas Reservoirs*, 2018 25(5): 56–60.

[25] Yuzhu Kang, Zhijiang Kang. Progress of geomechanics in hydrocarbon exploration in Tarim Basin. *Journal of Geomechanics*, 1995 1(2): 1–10.

[26] Siguang Li. *Introduction to Geomechanics*. Beijing: Science Press, 1973.

[27] Siguang Li. Vortex structure and other problems relating to the compounding of geotectonic systems of Northwestern China. *Acta Geologica Sinica*, 1954 34(4): 339–410.

[28] Hongzhen Wang, Sennan Yang, Benpei Liu. *Structural Paleogeography and Biogeography of China and its Adjacent Areas*. Wuhan: China University of Geosciences Press, 1990.

[29] Xia Zhu. *Structure and Evolution of Mesozoic-Cenozoic Basins in China*. Beijing: Science Press, 1983.

[30] Yuzhu Kang. The relationship between structural system and hydrocarbon in the Tarim Basin. In *Geomechanics*, Volume 9. Beijing: Geological Publishing House, 1989.

[31] Yuzhu Kang, Liusheng Ye, Zhihong Kang, et al. *Petroleum Geological Characteristics and Resource Evaluation in Northwest China*. Urumqi: Xinjiang Science and Technology Health Press, 1997.

[32] Yuzhu Kang, Zhenwei Gan, Zhihong Kang, et al. *Hydrocarbon Distribution and Exploration Experience in Major Basins of China*. Urumqi: Xinjiang Science and Technology Press, 2004.

[33] Yuzhu Kang. Establishment of China's theory of hydrocarbon geology. *Acta Geologica Sinica*, 2010 84(9): 1231–1274.

Features of Structural Systems in Northern China and their Control on Basin and Hydrocarbon Distribution

Abstract

In this study, the following are researched on the basis of structural system theory: basins, source rocks, hydrocarbon accumulation areas controlled by structural systems, hydrocarbon accumulation controlled by stresses, and oil fields controlled by subordinate shear structural systems. The results show that northern China is dominated by Neocathaysian, Western Region, zonal, Qi-Lv-He epsilon-shaped systems and other structural systems, and features related to migration, inheritance, discrepancy, transition, and complexity, which control the generation and development of the Tarim Basin, the Ordos Basin, and the Songliao Basin. The rules of petroliferous basins and hydrocarbon distribution as controlled by structural systems are explained, including superposition, conjunction, multi-order, and multi-stage control of structural systems and subordinate shear structural systems. The favorable areas for hydrocarbon accumulation are identified by utilizing structural system theory and considering the conditions of hydrocarbon accumulation comprehensively. Major breakthroughs and discoveries have been made in selected favorable areas, which prove the scientific reasoning and accuracy of the study on structural system control over oil and the prediction of favorable areas for hydrocarbon accumulation. These favorable areas are also the main targets for hydrocarbon exploration in the long term.

Keywords: Northern China, structural system, Tarim, Ordos, Songliao, Basin control, oil control.

1. Introduction

Guided by the theory of geomechanics, as established by Mr. Li Siguang [1–6], a series of oil fields, such as Daqing Oil Field, Dagang Oil Field, and Shengli Oil Field, has been successively discovered in China. Guided by the geomechanics [7–11], Kang Yuzhu achieved the first major breakthrough in a survey on Paleozoic Marine hydrocarbon in China in 1984 [12, 13], and discovered the world-class Tahe Oil Field in 1990 [14], followed by several hydrocarbon fields. At present, many scholars have carried out systematic research on extrabasinal structural systems, but there are few studies on intrabasinal structural systems, especially deep structural systems and their oil control applications.

Therefore, in this chapter, taking the north of China as the research object, where the structural systems (typical basins, such as Tarim Basin in the west, Ordos Basin in the middle and Songliao Basin in the east) are studied in depth, and based on new geological and geophysical data, a new understanding of petroleum geology, and new petroleum exploration achievements, the features and the basin control and oil control functions of the structural systems in petroliferous basins are systematically studied, and favorable hydrocarbon accumulation areas and exploration directions are indicated, so as to guide future hydrocarbon exploration.

2. Structural System Pattern

Over the years, in the theory of geomechanics, it has been recognized that there are four major structural systems in northern China, namely, Neocathaysian, Western Region, zonal, and epsilon-shaped structural systems [15–18], and the conjunction of these giant structures has resulted in several Paleozoic–Meso–Cenozoic petroliferous basins (Figure 1).

2.1. *Neocathaysian Structural System*

The Neocathaysian structural system was put forward by Professor Li Siguang in the 1930s and detailed in the early 1960s. The Neocathaysian

Figure 1. Main structural systems in northern China.

Notes: 1 — Zonal structural system; 2 — Meridional structural system; 3 — Cathaysian system; 4 — Neocathaysian system; 5 — Western Region system; 6 — Hexi system; 7 — Qinghai–Xizang–Sichuan–Yunnan inversed-S shaped structure; 8 — Epsilon-shaped and arc structures; 9 — Oil areas controlled by the Western Region structural system; 10 — Oil areas controlled by Neocathaysian structural system; 11 — Oil areas controlled by Pamir–Himalaya inversed-S shaped structural system.

system is a large-scale ξ-shaped system in eastern China and East Asia on rim of the Pacific. It is N-N-E trending (8°–25°) and mainly composed of three giant uplift zones and giant depression zones, which are referred to as "Three Uplifts & Three Depressions." These zones are roughly parallel to each other, as shown in Figure 2.

The first uplift zone consists of the Kuril Islands, the islands of Japan, the Ryukyu Islands, and the islands of the Philippines.

The first depression zone consists of the Okhotsk Sea, the Japan Sea, the Yellow Sea, the East China Sea, and the South China Sea.

The second uplift zone consists of the Zhugejuer Mountains, the Sikhote Mountains, the Zhangguangcai Mountains, the Laoye Mountains, the Changbai Mountains, the Wolf Forest Mountains, the Liaodong Peninsula, the Shandong Peninsula, and the southeastern coastal hills.

The second depression zone consists of the Songliao Basin, the North China Basin, the Jianghan Basin, and the Beibu Gulf Basin from north to south.

The third uplifting zone consists of the fold zones of the Greater Khingan Mountains, the Taihang Mountains, and the eastern part of the Guizhou Plateau.

The third depression zone consists of the Hailar Basin, the Erlian Basin, the Ordos Basin, and the Sichuan Basin from north to south.

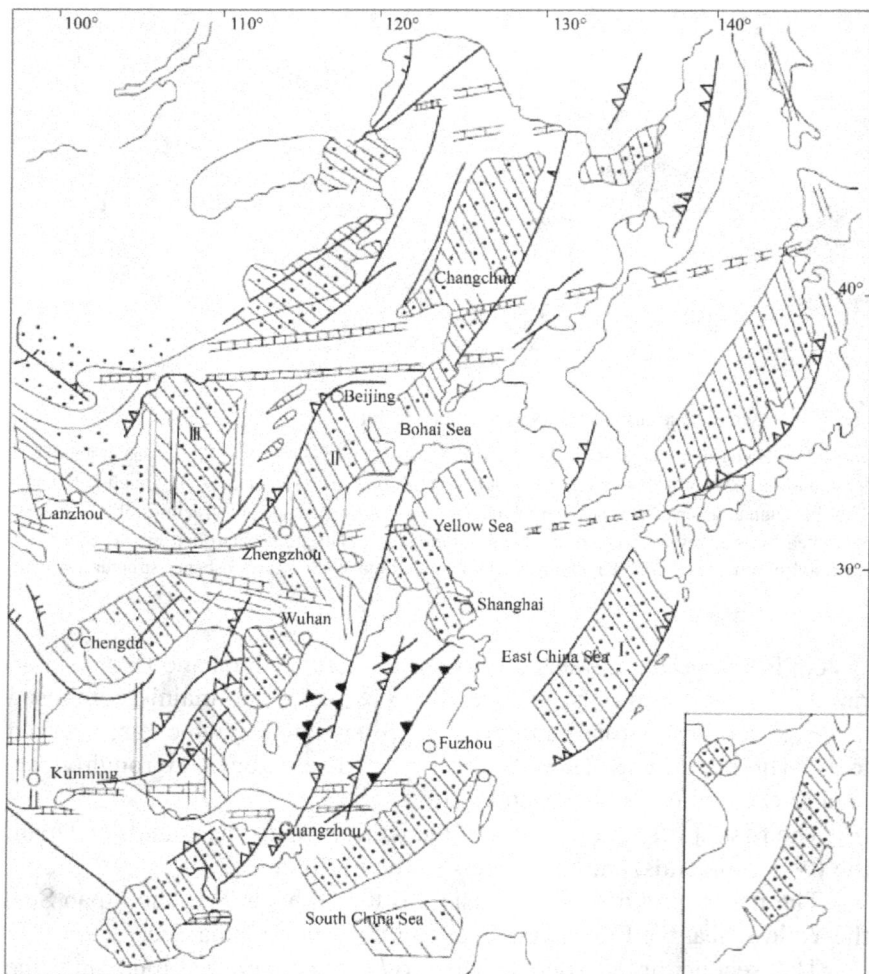

Figure 2.   Neocathaysian structural system.

Note: I — First depression zone; II — Second depression zone; III — Third depression zone.

The Neocathaysian structural system is mainly formed in the Meso–Cenozoic, and there are pre-Paleozoic, Paleozoic, and Meso–Cenozoic deposits and magmatic rocks from different periods. The Paleogene and Neogene strata in the first depression zone are packed and thick, and are characterized by marine deposits. The Meso–Cenozoic strata in the second and third depression zones are also packed. The Neocathaysian structural system was active earlier, mainly in the Meso–Cenozoic. In the process of its formation and development, volcanic activities were frequent and intense, including invasions and eruptions, and some parts are still active, especially the first uplift zone and the first depression zone.

2.2. *Western Region Structural System*

The Western Region structural system consists of a series of equally spaced and parallel N-W-W-trending (270°–310°) large conjunction structural zones. Each conjunction structural zone has a long history of formation and development, with formation in the Proterozoic and activities in the Paleozoic and even in the Meso–Cenozoic, dominated by uplift zones composed of large compressive-torsional faults and conjunction folds as well as a series of ξ-shaped depression zones (basins) on the south and west sides. These N-W-W-trending structural zones have experienced significant rightward torsional movement.

The formation and development of the Western Region system can be divided into two stages. In the first stage, there was a deep sea trough trending N-W-W in the Early Paleozoic, which was folded into mountains at the end of the late Silurian, forming a conjunction uplift structural zone with strong invasion and metamorphism. In the second stage, in the late Paleozoic, this developed into three pairs of large conjunction and alternate uplift and depression zones.

2.3. *Zonal Structural System*

There are two well-developed zonal structural systems in this area, namely, the Yinshan–Tianshan system and the Kunlun–Qinling system, which are very obvious in geomorphology and dominated by nearly E-W-trending folds and faults on the whole. The Yinshan–Tianshan zonal structural system is mainly located between 40°30′N and 42°30′N, traversing Xinjiang, Inner Mongolia, and Liaoning and Jilin provinces, and consisting of the Tianshan, Beishan, Yinshan, and Yanshan Mountains from west to east, with a length of more than 4,000 km. The Kunlun–Qinling zonal structural system spreads roughly between 32°30′N and 34°30′N, and consists of three sections, namely, the Kunlun Mountains in the west, the Qinling Mountains in the middle, and two branches in the east, one branch under the North China Plain and the Yellow Sea through

Songshan Mountain and the other branch consisting of the Funiu Mountain and Dabie Mountain structural zones, with a length of more than 4,000 km.

The zonal structural systems have a long history of development and multi-stage activities. They might have been active in the Archaean, and then experienced the Caledonian, Hercynian, Indosinian, Yanshanian, Himalayan and other multi-stage structural movements. They are still active today, with different development histories. For example, the Yinshan–Tianshan zonal structural system entered an important stage of development in the late Proterozoic, while the Kunlun–Qinling zonal structural system entered an important stage of development in the early Paleozoic. In each structural movement, the corresponding structural deformation and geological construction remained, forming complex structural zones.

2.4. *Qi-Lyu-He Epsilon-shaped Structural System*

The Qi-Lyu-He epsilon-shaped structural system is the largest one in China (Figure 3), consisting of the Qilian Mountains, the Taihang Mountains, the Luliang Mountains, the northern foot of the Qinling Mountains, and the Helan Mountains. The east and west wings of this system are asymmetrical, containing a series of ξ-shaped structural components. The front arc is mitered with the Qinling–Kunlun structural zone.

The Qi-Lyu-He epsilon-shaped structural system is located in central and northern China, spanning Xinjiang, Qinghai, Gansu, Ningxia, Shaanxi, Shanxi, Hebei, and Beijing, and other provinces, municipalities,

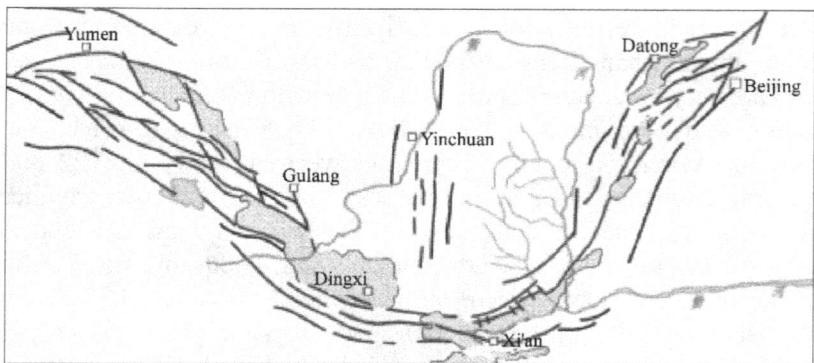

Figure 3.   Qi-Lyu-He epsilon-shaped structural system.

and autonomous regions, and sandwiched between the Tianshan–Yinshan zonal structural system and the Kunlun–Qinling zonal structural system, with the geological range of 92°00′E–120°00′E and 34°00′N–41°00′N, a length of 2,000 km from east to west, and a width of 900 km from north to south.

Its front arc overlaps the Kunlun–Qinling zonal structural system, and its arc top is located near Baoji. The west wing of the front arc extends along Jiuquan, Minle, Lanzhou, and Dingxi, showing an inversed ξ-shaped arrangement of N-W-trending fold zones, fault zones, and troughs in between. The east wing of the front arc extends along Hancheng, Lishi, Ningwu, and Datong, showing a ξ-shaped arrangement of N-E-trending large anticlines and synclines.

Its spine consists of the Helan Mountain fold zone, which is consistent with the Helan Mountain meridional structural zone; that is, the Helan Mountain fold zone overlaps the Helan Mountain meridional structural system.

The shields of this system are located on the east and west sides of the spine, with the Aning Shield on the west side and the Yishan Shield on the east side.

The west wing of the reflection arc spreads along Jiuquan, Yumen, Subei, Annanba, and other places, and is an arc-shaped structural zone protruding to the north, with the arc top in the north of Qiaowan. The east wing of the reflection arc spreads along Datong, Xuanhua, Chengde, Qinhuangdao, and other places, and is an arc-shaped structural zone protruding to the north, with the arc top near Chengde.

3. Features of Structural Systems

All structural systems develop some common features in the process of formation and development, which can be summarized as follows: stage, migration, inheritance, differences, transition, and complexity.

3.1. *Stage*

The evolution of the structural system is not uniform, showing a stage feature at times.

The Tarim Basin consists of the Shaya paleo-uplift and the Kuruketagh Uplift in the zonal system. It was an E-W-trending depression zone in the

Sinian–Middle Ordovician, uplifted initially in the Middle and late Ordovician, and was rapidly uplifted and formed in the Silurian–Devonian and even in the Carboniferous–Permian. Some areas suffered denudation, and depression in the Meso–Cenozoic, and most areas were covered.

The central paleo-uplift in the Ordos Basin emerged at the end of the Cambrian, entered the development peak in the Middle and Late Ordovician, was exposed for a long time in the Silurian and Devonian, and entered a continuous burial period in the Carboniferous. In the Late Triassic, influenced by the development of the foreland basin in the western margin of the Ordos Basin, the high point of the paleo-uplift migrated to the east and arrived at the east of the Ordos Basin in the Late Cretaceous, and the original high point of the uplift became a depression.

On the basis of the regional uplift in the late Permian–Triassic, the central paleo-uplift in the Early Middle Jurassic and Early Cretaceous strongly extended on both the east and west sides and formed the central paleo-uplift of the Songliao Basin. During the deposition period from the Late Jurassic Huoshiling Formation to the Early Cretaceous Dengluoku Formation, the high point of the central paleo-uplift experienced weathering and denudation for 20 Ma, was covered by the Dengluoku Formation and the above strata in the middle-Late Cretaceous, and then entered a stage of continuous and stable burial.

3.2. *Migration*

The evolution of the structural system was not uniform, showing a migration feature (fault, rock slurry movement, sedimentation center, depression center, etc.).

The depression center of the Tarim Basin in the early Paleozoic was located in the Mangar area of the Western Region system, and migrated to the Yecheng area in the late Paleozoic. The depression center of the basin in the Triassic–Jurassic was located at the Kuqa Depression of the Tianshan zonal system, and migrated to the Kashi Depression of the Kunlun zonal system in the Cenozoic.

Since the Cambrian, the deposition centers in the Ordos Basin have migrated to different degrees in different periods; for example, the basin deposition center was located in the southwest in the Ordovician and migrated to the west and east in the Carboniferous.

3.3. *Inheritance*

The structural system in a certain geological period is based on the structural system in the previous geological period.

There are at least two inheritance developments in the Neocathaysian system and the Qi-Lyu-He epsilon-shaped structural system, and four structural system in the zonal system and Western Region system in northern China, such as at the end of the Middle Ordovician, at the end of the Silurian, at the end of the Devonian, and at the end of the Carboniferous–Permian. Under the multi-stage subordinate regional structural movements, there were inheritance activities in the abovementioned structural systems with different features.

The Tazhong Uplift in the Tarim Basin emerged in the middle and late Middle Caledonian Period, and was basically formed in the late Caledonian and early Hercynian Periods. After the formation of the uplift, it was maintained in a stable structural environment for a considerable period, with small structural changes, relatively complete strata, and micro denudation in the local parts; the later structural movement was dominated by full uplift and depression (in the Meso–Cenozoic or Late Paleozoic).

The Yimeng Uplift in the zonal system of the Ordos Basin is an E-W-trending structure that inherited the form of a crystalline basement, and the Taiyuan Formation in the Upper Carboniferous directly overlays the metamorphic basement.

The Changling Conjunction Fault Depression group in the Songliao Basin, controlled by the Neocathaysian system, experienced three stages, namely, initial rifting in the sedimentary period of the Huoshiling Formation, intense rifting in the sedimentary period of the Shahezi Formation, and rifting shrinkage in the sedimentary period of the Yingcheng Formation. Fracture succession activities at the main trunk boundary of the control trap always affect the trap's deposition, leading to the superposition of the entanglement in different periods [19].

3.4. *Differences*

The activity features of a structural system in a certain geological period are different from those in the preceding and subsequent structural periods, and they are also different at different parts of the same structural system.

The Tarim Basin belongs to the Luntai Fault in the zonal system. During its formation and evolution process, it experienced many activities of different intensities and natures in different stages of the fault; therefore, these different areas have different control effects on hydrocarbon formation. The eastern part of the fault was uplifted higher, while the western part was uplifted lower, resulting in the distribution in the Upper Pangeozoic and Neoproterozoic being old in the east and new in the west and the different strata in the Mesozoic being directly overlaid on those formed in the Paleozoic.

There are obvious differences in fault activity features between the spine of Qi-Lyu-He epsilon-shaped structural system and the northern, central, and southern parts of the western margin of the Ordos Basin structural zone. The northern part is a thrust system, in which a series of westward overthrust faults are developed in the northern section, a simple large overthrust uplift and front triangle zones are developed in the middle section, and many rows of eastward overthrust faults are developed in the southern section. The middle part is the nappe system, a typical thin-skinned structure is developed in the northern section, a thick-skinned structure is developed in the middle section, and a large-angle imbricated structure is developed in the southern section. The southern part is an overthrust system with a large-angle westward thrust.

The formation mechanism of the Songliao Basin located in the second depression zone of the Neocathaysian system is obviously different from that of the Tarim Basin and the Ordos Basin. The Songliao Basin was formed in a strong tensile environment, the Tarim Basin was formed in a strong compressional environment, and the Ordos Basin was formed in a weak compression, weak tension environment. The strong tensile background of the Songliao Basin leads to the entry of mantle-derived gas into the basin along deep faults or volcanic channels leading to hydrocarbon formation, and many carbon dioxide gas reservoirs, helium anomalies, and inorganic hydrocarbon gas have been found.

3.5. *Transition*

The transition of the structural system due to compression and tensile stress, rise and subsidence, and uplift and depression in different geological periods is an important feature.

The Shaya paleo-uplift in the E-W-trending structural zone in the northern part of the Tarim Basin was a depression zone in the early

Paleozoic, which received deposits in the Cambrian and Ordovician, and was transformed into an uplift zone in the Silurian and Devonian. The boundary (Yanan) fault in the northern margin of the Shaya paleo-uplift was a southward compression-torsional fault zone before the Himalayan period and was transformed into a southward tension fault zone during the Himalayan period.

There are multiple transitions in the stress field in the Ordos Basin. In the early and Middle Jurassic, the basin was in a tensile stress environment. From the Late Jurassic onward, the structural stress field was transformed into a compression field, and the rim of the basin was subjected to multiple compressive stresses. In the early Cretaceous, the structural stress field of the basin was transformed into the tensile stress field. From the late Early Cretaceous to the Late Cretaceous, the structure of the basin was transformed from the previous tensile stress into compressive stress.

There were many transitions in the stress field in the Songliao Basin controlled by the Neocaysian system. It was in a compressive stress environment in the pre-rift period, in the N-W-trending stress field in the fault depression period, in the nearly E-W-trending tensile tress field in the depression period, and then in the N-N-W-trending to S-S-E-trending compressive stress field in the reversal period.

3.6. *Complexity*

The formation and development of structural systems often occur over a long geological period, resulting in multiple types of structural systems, multiple orders and grades of structural patterns, different structural types of oil control, and various oil control functions in different parts of the same structure. For example, in the Tazhong area of the Tarim Basin, five groups of faults are developed, namely, the N-W-trending, N-E-trending, N-E-E-trending, nearly E-W-trending, and nearly N-S trending faults, among which, the N-W-trending faults dominate, with a long extension and large fault spacing. There are four main periods of fault activities, namely, the middle Caledonian, the late Caledonian, the Hercynian–Indosinian, and the Himalayan. The activity initiation period, times, and main activity periods of each fault are different from those of other faults. The faults in different developmental stages (syn-sedimentation fault, post-sedimentation fault, hydrocarbon generation fault, and post-accumulation fault) have different control effects on hydrocarbon [20].

4. Basin Control of Structural Systems

Many years of research and practice have shown that the formation and development of the main structural systems control the formation and development of the oil–gas basins (Table 1) [11, 15–18].

The Tarim area was mainly controlled by the zonal system, the Western Region system, and their conjunctions in the Paleozoic; its sedimentary construction was mainly controlled by the zonal system before the Middle Ordovician, by the zonal system (dominated) and Western Region system from the Middle Ordovician to the Devonian, and by the Western Region system in the Carboniferous–Permian. Since the Meso–Cenozoic, the Tarim Basin has mainly been controlled by the conjunction of the zonal system, the Hexi system, the southern Tianshan arc, the Hetian arc, the Pamir ξ-shaped structure, and the Qinghai–Xizang ξ-shaped structure. It has experienced six evolution stages, namely, the rift craton basin in the Sinian–Middle Ordovician, the flexural basin in the Siluran–Devonian, the craton depression basin in the Carboniferous–Permian, the fault basin in the Triassic–Jurassic, the fault depression basin in the Cretaceous–Paleogene, and the intracontinental unified basin in the Neogene–Quaternary, forming the present basin structural pattern [15, 16].

The Ordos Basin was mainly controlled by the epsilon-shaped structure, the zonal system, the Neocathaysian system, the Cathaysian system, and the meridional system. It was mainly controlled by the Cathaysian system, the zonal system, and Helan epsilon-shaped system from the Paleozoic to the Triassic, and by the Neocathaysian system, the zonal system, and the Helan epsilon-shaped system after the Triassic. It has experienced five evolution stages, namely, the intracraton rifted basin stage in the Meso–Proterozoic, the epicontinental sea stage in the Sinian–Early Paleozoic, the intracratonic depression basin stage in the Late Paleozoic–Early Mesozoic, the foreland basin stage in the Meso–late Mesozoic, and the fault depression basin stage in the Cenozoic [17].

The Songliao Basin was mainly controlled by the zonal system, the Neocathaysian system, the Cathaysian system, and the meridional system. The evolution of the basin was jointly controlled by the zonal structural system and the Cathaysian structural system in the Cambrian–Devonian, by the zonal structural system and the Neocathaysian structural system in the Carboniferous–Permian, and by the Neocathaysian system in the Meso–Cenozoic. It has experienced five evolutionary stages, namely, the

Table 1. Relationship between petroliferous basins and structural systems in northern China.

Basin	Structural system						
	Zonal system	Western Region system	Cathaysian system	Neocaysian system	Meridional system	Inversed-S-shaped system	Epsilon-shaped system
Tarim	✓	✓				✓	
Ordos	✓		✓	✓	✓		✓
Songliao	✓		✓	✓			

rift craton basin stage, the craton-like basin stage, the craton depression basin stage, the fault depression basin stage, and the depression basin stage [18].

5. Oil Control Distribution Law of Structural Systems

Based on the theory of geomechanics and the structural systems, the control of structural systems on the basins, the hydrocarbon source areas, and hydrocarbon accumulation zones; the control of stress on hydrocarbon fields; and the control of low-order twisting structures on hydrocarbon fields are studied to obtain the hydrocarbon control distribution law of structural systems.

5.1. *Multi-stage Control of Hydrocarbon by Structural Systems*

(1) **Compound control of petroliferous basins by giant structural system:** Various large petroliferous basins have been formed by conjunction of the zonal system, the Western Region system, the Cathaysian tectonic system, the Neocathaysian structural system, and other structural systems.

(2) **Formation of basins controlled by the Neocathaysian system:** Since the Meso–Cenozoic, a series of large and medium-sized petroliferous basins have been formed mainly by control of the Neocathaysian system in central and eastern China, among which the Ordos Basin and the Songliao Basin are located in the third and second depression zones of the Neocathaysian system, respectively.

(3) **Formation of basins controlled by the Western Region system:** Since the late Paleozoic, a series of large and medium-sized petroliferous basins have been formed by control of the Western Region structural system in conjunction with the zonal structural system in northwest China, including the Tarim Basin.

(4) **Hydrocarbon source areas controlled by first-order sedimentation zones (depressions):** The first-order depression zones (depressions) controlled by major structural systems are generally the main hydrocarbon source areas, in which there are multi-stage and multi-layer hydrocarbon source rocks, such as in the Tarim Basin, the Ordos Basin, and the Songliao Basin (Table 2).

Table 2.   Structural systems in northern China controlling hydrocarbon.

Basin	First-order hydrocarbon source area (depression)	First-order hydrocarbon enrichment zone (uplift and slope)
Tarim	Kuqa Depression (E-W), A-Man Depression (N-W), Kashi–Yecheng Depression (N-W, inversed-S, E-W), Jiemu Fault Depression (inversed-S)	Kuqa Foreland Fault Fold Zone–Shaya Uplift (E-W), Bachu Uplift–Katak Uplift (N-W), Gucheng Ruins (N-E), Maigeiti Slope (N-W), Kongquhe Slope (N-W)
Ordos	Yishan Slop–Tianhuan Depression (S-N), Weihe Depression (E-W)	Yishan Slope (S-N), Western Region Overthrust Zone (S-N), Weibei Uplift (E-W)
Songliao	Central Depression (N-N-E)	Daqing Long Wall and Western Slope (N-N-E)

In the Tarim Basin, the Kuqa Depression was controlled by the zonal system, the Awatti–Mangar Depression was controlled by the Western Region and zonal systems, and the Kashi–Yecheng Depression was controlled by the zonal system, the Western Region system, and the Pamir inversed-S-shaped structural system, in which Lower Paleozoic, Upper Paleozoic, and Mesozoic source rocks developed, forming major hydrocarbon source areas.

The southern and northern parts of the Ordos Basin were mainly controlled by the zonal structural system. The western part was mainly controlled by the conjunction spine of the meridional structural system and the Qi-Lyu-He epsilon-shaped structural system. The central part was located in the shield on the east side of the Qi-Lyu-He epsilon-shaped structural system and mainly controlled by the Cathaysian and the Neocathaysian structural systems (Figures 4 and 5).

The Songliao Basin is located in the second depression zone of the Neocaysian system. Under the control of the Neocaysian system, several fault basins were formed in the Jurassic–Early Cretaceous. Each fault depression had its own independent source, reservoir, and cap, forming its own independent petroliferous system. The Upper Cretaceous system is the main source layer of the Daqing Oil Field, and hydrocarbon is mainly distributed in the anticline zone in the central depression area of the Songliao Basin (commonly known as the Daqing Long Wall).

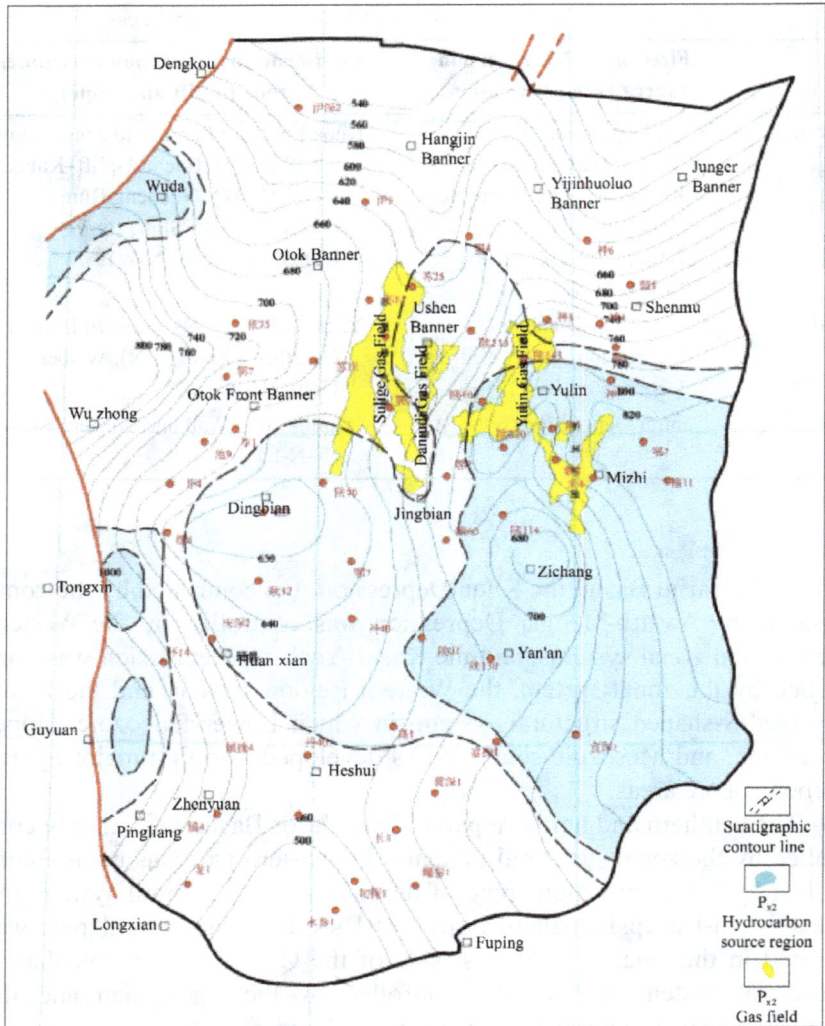

Figure 4.   Relationship between upper Paleozoic gas field and source rock in the Ordos Basin.

In the central and western basins developed against the background of regional compression stress, the hydrocarbon source areas are mostly composed of multiple inversed structural systems, and the hydrocarbon source rocks are characterized by multilayers, large

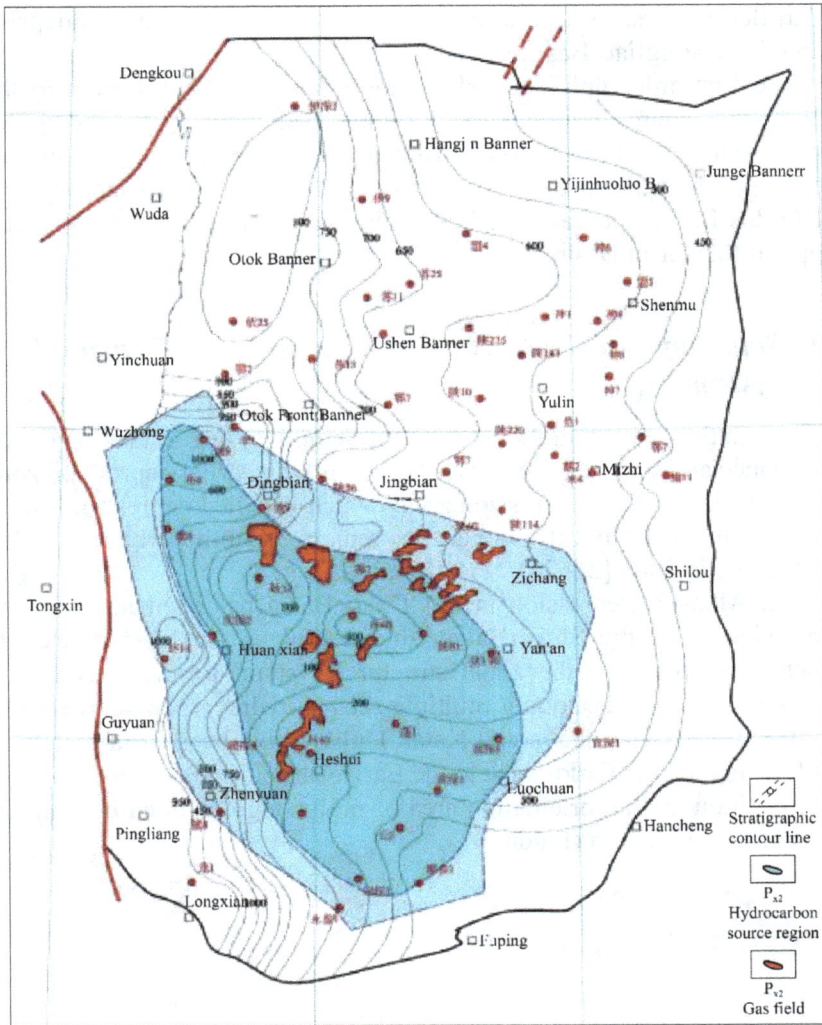

Figure 5. Relationship between Mesozoic oil field and source rock in the Ordos Basin.

thickness, and wide distribution in plane, such as the western and southern depressions of the Ordos Basin and the southwest and north depressions of the Tarim Basin. In the basins in eastern China developed under regional tensile stress, the hydrocarbon source areas are controlled by first-order depression zones, such as the central depression of the Songliao Basin.

The first-order uplift and slope zones immediately adjacent to the hydrocarbon source areas and the uplift and fault zones within the hydrocarbon source areas first become hydrocarbon accumulation areas, such as the Daqing Long Wall of the Songliao Basin, the central paleo-uplift of the Ordos Basin, the Tazhong Uplift, the Shaya Uplift, and the Maigeti Slope of the Tarim Basin.

5.2. *Superimposed Control of Hydrocarbon by the Structural Systems*

The formation and evolution of the structural systems have involved inheritance and stages. In geological structures, an earlier depression zone continues to remain a depression zone, which is called depression superposition; an earlier uplift continues to remain an uplift, which is called uplift superposition [21–24].

The Mangar Depression in the Tarim Basin has been in a state of depression since the Sinian under the control of the Western Region structural system and the zonal structural system, forming depression superposition and developing multiple sets of hydrocarbon source rocks ($\mathcal{E}$-O, $\mathcal{E}$-P_1, T). The Shaya and Katak Uplifts have been in in a state of uplift from the Late Ordovician to the Early Carboniferous, forming uplift superposition and an orientation area of multi-stage hydrocarbon migration and accumulation (Figure 6).

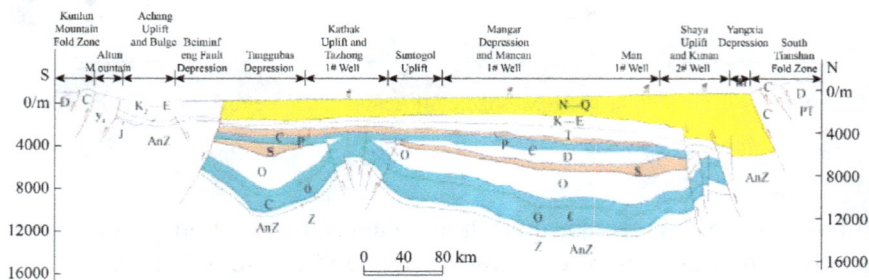

Figure 6.   Tazhong 1# well, Mancan 1# well, and Kunan 2# well in the Tarim Basin.

The Ordos Basin has been in a state of depression under the control of the Cathaysian structural system, the Helan epsilon-shaped structural system, the zonal structural system, and the Neocaysian structural system since the Cambrian, forming depression superposition, developing multiple sets of hydrocarbon source rocks (Є-O, C-P, T-J), and becoming a large hydrocarbon source area. The paleo-uplifts in the Ordos Basin, such as the Yimeng Uplift and the central paleo-uplift, have strong inheritance, and the uplift locations basically remain unchanged, forming major hydrocarbon accumulation zones, many large gas fields, such as in Jingbian, Sulige, Hangjin Banner, Danudi, Yulin, and large oil fields, such as in Jiyuan and Qingyang.

The Songliao Basin has been in a state of depression under the control of the Neocathaysian structural system, the zonal structural system, and the Cathaysian structural system since the Upper Paleozoic, developing Carboniferous–Permian and Jurassic–Cretaceous hydrocarbon source rocks and forming large hydrocarbon source areas. The central uplift zone and the surrounding slope zone in the basin have become the main hydrocarbon accumulation zones, and a series of large and medium-sized hydrocarbon fields, such as the one in Daqing, have been discovered.

5.3. Compound Control of Hydrocarbon by the Structural System

In northern China, the composite and joint control of oil- and gas-bearing basins by giant tectonic systems is evident.

The A-Man Depression in the Tarim Basin is the result of compound effect of the zonal system and the Western Region system, and the Kashi–Yecheng Depression is the result of compound effect of the Western Region system, the zonal system, and the Qinghai–Xizang inversed-S-shaped system.

In the Songliao Basin, the evolution of the Cambrian–Devonian basins was controlled by the conjunction of the zonal structural system and the Cathaysian structural systems, and the evolution of the Carboniferous–Permian and Meso–Cenozoic basins was controlled by the conjunction of the zonal structural system and the Neocathaysian structural system.

In the Ordos Basin, the west part was mainly controlled by the conjunction spine of the meridional structural system and the epsilon-shaped structural system, and the central part was mainly controlled by the conjunction of the Cathaysian structural system, the Neocathaysian structural system, and the regional E-W-trending structural zones.

5.4. *Multi-stage Control of Hydrocarbon by Structural Systems*

Each structural system has a very long activity period. For example, the zonal system has been active since the Archaean and the Western Region system has been active since the late Ordovician. In this long geological period, the activity intensity of structural systems is not uniform, showing a multi-stage feature, which has resulted in multi-stage features of hydrocarbon accumulation.

There have been four hydrocarbon accumulation periods in the Tarim Basin since the Early Paleozoic, namely, two primary hydrocarbon accumulation periods in the middle Caledonian and the Indosinian, and two adjustment hydrocarbon accumulation periods in the middle Yanshan (J_3-K_1) and the Late Himalaya (N_2-Q).

There have been two hydrocarbon accumulation periods in the Ordos Basin, namely, the Indosinian and the Yanshanian. The Mesozoic and Cenozoic Ordos Basin is characterized by a stable structure and gentle strata (at the dip angle of only about 1° westward), resulting in a long and continuous accumulation feature.

There have been two hydrocarbon accumulation periods in the Songliao Basin, namely, the Yanshanian and the Himalayan. Due to the special geological background, a carbon dioxide gas reservoir was developed and formed in the Songliao Basin in the Himalayan period, which was later than the accumulation of hydrocarbon gas.

5.5. *Distribution of Hydrocarbon Fields by the Subordinate Shear Structural Systems*

According to Mr. Li Siguang, "The regions belonging to the same parts of the same type of structural system have a great consistency in deposition and structure conditions, although there are some differences. In other words, in the same oil region, all oil fields have common features in the structure, in addition to their own features. If oil is found in one of the structures, it may be found in other adjoining structures." Mr. Li Siguang also stated that the core of hydrocarbon control is the subordinate shear structural system, which is also the key to studying the hydrocarbon distribution law.

Based on many years of practice, six oil control models of hydrocarbon using the subordinate shear structural system have been established, namely, broom-shaped, echelon, rotation-torsional, inversed-S-shaped, λ-shaped, and imbricated models [25]. Hydrocarbon controlled by the broom-shaped structure is mostly concentrated in the middle part and at the dispersal parts, showing a broom shape on the whole. The enrichment degree of hydrocarbon in the rotation-torsional structure becomes worse from the inner cycle layer to the outer cycle layer. Hydrocarbon controlled by the inverted-S-shaped structure is mainly distributed in the arc-shaped zone with moderate stress. Hydrocarbon controlled by the λ-shaped structure is usually distributed at locations of the acute angle between the main fault and the branch fault. Hydrocarbon controlled by the echelon structure and imbricated structure is also distributed in the echelon and imbricate forms, respectively.

6. Favorable Hydrocarbon Accumulation Aareas and Exploration Directions

Based on the abovementioned studies, taking into account the conditions of hydrocarbon accumulation and oil control by structural systems, the prospect of hydrocarbon resources has been evaluated, and the favorable areas and exploration directions for hydrocarbon accumulation are indicated [15–18].

6.1. *Favorable Area in the Tarim Basin*

Favorable areas for hydrocarbon accumulation in the Tarim Basin [15, 16] are as follows: the Shaya Uplift, the Kuqa Depression, the Katak Uplift, the Maigeti Slope, the southern slope of the Yecheng Depression, the Wushi Depression, the Caohu Slope, the Yangxia Uplift, the Atushi Anticline Zone, the southern fault of the Bachu Uplift, the western part of the Tazhong Uplift, the southern slope of the Tazhong Uplift, the footwall of the Tazhong Fault, the northern slope of Shuntoguole, the northern and southern slopes of the Bachu Uplift, the northern slope of the Manchu Uplift, the nose bulge of the Koerla Uplift, the Tanggubas Depression, the Shache Low Bulge, the Moyu Bulge, and the southern slope of the Gucheng Ruins Uplift.

6.2. *Favorable Areas in the Ordos Basin*

Gas is distributed in the whole of the Ordos Basin and oil is distributed in the southern part of the Ordos Basin [17]. The favorable areas of the gas accumulation are the Sulige–Hangjin Banner region in the north, the Jingbian–Yan 'an region in the middle, the Zizhou–Mizhi region in the east, the Yanxia regions in the northwest and east, the Huating region in the southwest, and the Fuxian–Yichuan region in the southeast. Favorable areas for oil accumulation are northern Shaanxi, the area surrounding the Xifeng Oil Field, Jiyuan, Huaqing, the fault fold zone on the western margin, and the southeast and southwest parts.

6.3. *Favorable Areas and Exploration Directions in the Songliao Basin*

Favorable areas and exploration directions for hydrocarbon accumulation in the Songliao Basin are as follows [18]: (1) Mesozoic-Cenozoic: fault zones, depression and slope areas, deep volcanic gas and shallow biogas; (2) Carboniferous–Permian: Zhaoyuan, Da'an-Nongan, Hai Tuozi–Changling, West Tongliao, Binnan, Qian'an–Changling, Manghan, and Changtu–Lishu.

Major breakthroughs and discoveries of new hydrocarbon accumulation zones have been made in selected favorable areas through exploration and prediction, which have proved the scientific reasoning and accuracy of oil control research of the structural system [15]. These favorable areas and exploration directions are also the main targets for hydrocarbon exploration in the long term.

References

[1]   Lee, J. S. *Introduction to Geomechanics*. Beijing: Science Press, 1973.

[2]   Researching and Mapping Group of Geomechanics in CAGS. *Specification of the Structural system Map of the People's Republic of China (1:400,000)*. Beijing: Geological Publishing House, 1978.

[3]   Institute of Geomechanics, Chinese Academy of Geological Sciences. *Specification of the Structural system <ap of the People's Republic of China and its Adjacent Sea Area (1:250,000)*. Beijing: Sinomaps, 1984.

[4]   Shujing Li, Daxing Zheng, Jiamu Chen, *et al. Division and Characterization of Main Structural systems in China/Collection of Research on China's Provincial Structural System (1)*. Beijing: Geological Publishing House, 1985.

[5] Dianqing Sun, Naigong Deng. Prospect of oil resources in China as viewed from geomechanics. *Bulletin of the Chinese Academy of Geological Sciences*, 1979 1(1): 59–66.

[6] Dianqing Sun. *Geomechanics Theory and Practice in Petroleum Prospecting in China. Bulletin of the Institute of Geomechanics, CAGS (13)*. Beijing: Geological Publishing House, 1989.

[7] Yuzhu Kang. *The Main Structural systems and Petroleum Distribution in China*. Urumqi: Xinjiang Science and Technology Press, 1999.

[8] Yuzhu Kang. The main structural systems and their controls on petroleum resource in China. *West China Petroleum Geosciences*, 2007 3(1): 1–8.

[9] Yuzhu Kang, Shuwen Xing, Yinsheng Ma, *et al. Introduction to the Global Structural System*. Beijing: China Petrochemical Press, 2018.

[10] Yuzhu Kang. Establishment of China's theory of hydrocarbon geology. *Acta Geologica Sinica*, 2010 84(9): 1231–1274.

[11] Yuzhu Kang, Yue Zhao, Yinsheng Ma, *et al. Petroleum Geomechanics*. Beijing: Geological Publishing House, 2012.

[12] Yuzhu Kang. The discovery of high-yielding oil flow in well Shashen 2 and the direction of oil prospecting in the future. *Oil & Gas Geology*, 1985 6(S1): 45–46.

[13] Yuzhu Kang, Zhijiang Kang. Progress of geomechanics in hydrocarbon exploration in Tarim basin. *Journal of Geomechanics*, 1995 1(2): 1–10.

[14] Yuzhu Kang, Ximing Zhang, Zhihong Kang, *et al. Tarim Basin and Tahe Oil Field in China*. Urumqi: Xinjiang Science and Technology Press, 2004.

[15] Yuzhu Kang, Zongxiu Wang, Xingui Zhou, *et al. Structural systems and Their Controls on Petroleum Resource in Northwest China*. Beijing: China Land Publishing House, 2013.

[16] Yuzhu Kang, Zongxiu Wang, Xiaofeng Wang, *et al. Structural Systems and Their Controls on Petroleum Resource in Tarim Basin*. Beijing: China Land Publishing House, 2009.

[17] Yuzhu Kang, Zongxiu Wang, Xingui Zhou, *et al. Structural systems and Their Control on Petroleum Resource in Ordos Basin*. Beijing: Geological Publishing House, 2014.

[18] Yuzhu Kang, Zongxiu Wang, Huijun Li, *et al. Structural Systems and Their Controls on Petroleum Resource in Songliao Basin*. Beijing: Geological Publishing House, 2015.

[19] Jianbao Yun, Zhijun Jin, Jingen Yin. Features of inherited fault zones and their effect on hydrocarbon accumulation. *Geostructurala et Metallogenia*, 2002 26(4): 379–385.

[20] Zhongpei Zhang, Yi Wang, Jinbiao Yun, *et al.* Control of faults at different evolution stages on hydrocarbon accumulation in Tazhong area, the Tarim Basin. *Oil & Gas Geology*, 2009 30(3): 316–323.

[21] Dongsheng Sun, Shuangjian Li, Jinbiao Yun, *et al.* The activities of paleo-uplifts and distribution of hydrocarbon in marine Craton Basins, China. *Acta Geologica Sinica*, 2017 91(7): 1589–1603.

[22] Dengfa He, Desheng Li, Xiaoguang Tong, *et al.* Accumulation and distribution of hydrocarbon controlled by paleo-uplift in poly-history superimposed Basin. *Acta Petrolei Sinica*, 2008 29(4): 475–488.

[23] Guoai Xie, Qinglong Zhang, Mingbao Pan, *et al.* Two different genetic types of paleouplift in the Ordos Basin and its significance in hydrocarbon exploration. *Geological Bulletin of China*, 2005 24(4): 373–377.

[24] Zhihong Kang, Liling Wei, Beichen Hu. Features of oil-gas formation by superposition of prototype Basins of Tarim. *Xinjiang Geology*, 2002 20(1): 58–61.

[25] Yuzhu Kang. A study on oil-control model of subordinate shear structural system in China. *Journal of Geomechanics*, 2018 24(6): 737–747.

A Study on an Oil Control Model of a Subordinate Shear Structural System in China

Abstract

Guided by the theory of geomechanics, the oil control effect of 8 (regional) structural systems, in the Tarim Basin, the Junggar–Tuha Basin, the Qaidam Basin, the Corridor Region, the Ordos Basin, the Sichuan Basin, the Songliao Basin, and Bohai Gulf, is studied. It is recognized that the distribution of oil and gas fields is mainly controlled by various subordinate shear structural systems. On this basis, the oil control model of five kinds subordinate shear structural systems is developed and established, including the broom-shaped structural system, the shear structural system, the echelon-shaped structural system, the λ-shaped structural system, and the inverted-S-shaped structural system. The formation characteristics, hydrocarbon accumulation conditions, and distribution rules of various subordinate shear structural systems are summarized with typical examples. It is believed that the formation of different structural systems is controlled by different structural environments and stress fields. The distribution rules of oil and gas fields are controlled by the different subordinate shear structures. The establishment of these oil control models is of great significance for the exploration and discovery of current and future oil and gas sources.

Keywords: Subordinate shear, structural system, oil control model, geomechanics theory.

Several decades of oil and gas exploratory practices have proved that the theory of geomechanics as established by Mr. Li Siguang is of great significance [1–4]. Under the guidance of this theory, the first major breakthrough in Paleozoic marine oil and gas sources was made in China in 1984, when the world-class Tahe Oil Field was discovered. In 1990 and later, several oil and gas fields were discovered [5–9]. In 2007, the oil control effects of structural systems were studied successively in the Tarim Basin, the Junggar–Tuha Basin, the Qaidam Basin, the Corridor Region, the Ordos Basin, the Sichuan Basin, the Songliao Basin, and Bohai Bay, and it was found that the distribution of oil and gas fields was mainly controlled by subordinate shear structural systems, including typical broom-shaped structural systems, shear structural systems, echelon-shaped structural systems, λ-shaped structural systems, as well as S-shaped and inverted-S-shaped structural systems [10–12]. In this paper, the formation characteristics, oil–gas accumulation conditions, and distribution rules of these five kinds of subordinate shear structural systems are described, and the oil control theory of structural systems is established, which has important significance for research on the current and future geomechanics theory and oil and gas explorations.

1. Oil and Gas Control of the Broom-shaped Structural System

1.1. *Oil and Gas Control of the Yakela Broom-shaped Structural System*

The Yakela broom-shaped structural system is located in the north midsection of the Shaya Uplift in the northern region of the Tarim Basin, connected to the Kuqa Depression by the Yanan Fault in the north and bounded by the Luntai Fault in the south, with the broom shape spreading to the southwest and converging to the east [13, 14], spanning an area of over 1×10^4 km^2 (see Figure 1).

1.1.1. *Features of the structural system*

Based on the analysis of the paleo-structural evolution of the fault, the Yanan and Luntai Faults have been active for multiple periods since the Paleozoic, and have included four major activity periods.

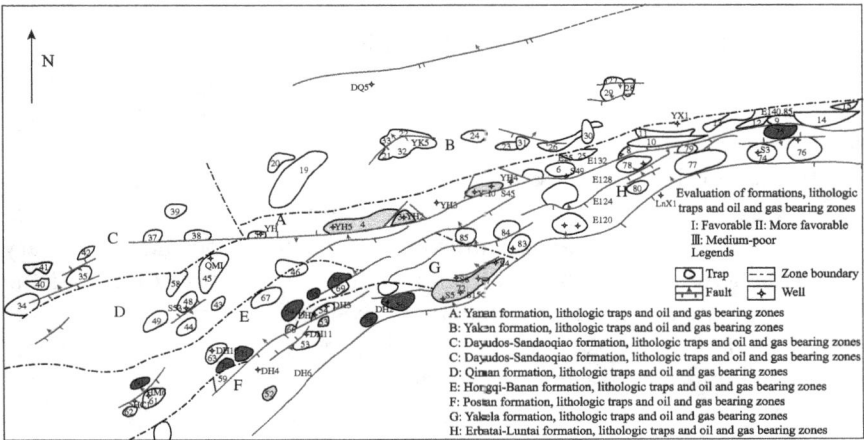

Figure 1. Distribution of oil and gas fields controlled by the Yakela broom-shaped structure.

In the Caledonian–late Hercynian Period, the two faults were active; and the thickness on both sides of the Luntai Fault changed significantly in the Paleozoic Period, resulting in huge fault spacing and displacement, making the northern section of the fault overthrust upward and become denudated.

In the Indosinian–Early Yanshan Period, these faults were continuously active and the eastern, middle, and western sections of the faults inherited the development in the Caledonian–late Hercynian Period. The main fault extended upward along the original position, the northward dip of the fault was overthrust, and the Luntai Fault decreased in the south and rose in the north. The Triassic–Jurassic System in the southern section had a thickness of 500–700 m, while there was almost no deposition of the Triassic–Jurassic System in the northern section aside from some minor deposition with a thickness of less than 200 m.

In the late Yanshanian–Early Himalaya Period (Cretaceo–Paleogene), the eastern section still showed an overthrust nature, with the section dipping northward; the Luntai Fault rose in the north and decreased in the south. The Cretaceous–Paleogene System had a thickness of 900–1,300 m in the southern section and 300–450 m in the northern section, while the faults in the middle and western sections were re-active and slipped, showing positive faults, with the section dipping northward; the northern section decreased and the southern section rose. There was little difference in the thickness of the two sections in the Chalk–Paleogene Period, with a thickness of 800–900 m in the northern section and 600–750 m in the southern section, indicating minor activities in the middle and western sections.

In the late Himalaya Period (after the Miocene), the eastern section showed a normal fault nature with the northern section slipping, the section dipping northward, and the thickness of the Neogene and Quaternary Systems increasing up to 5,000 m. The southern section rose, and the thickness of the Neogene and Quaternary Systems was up to 4,000 m. The middle and western sections were overthrust. In the late Himalayan Period, the middle and western sections only showed basement faults, but the Cretaceous–Neogene System was not faulted.

1.1.2. *Oil control of the Yakela broom-shaped structural system*

The main body of the Yakela broom-shaped structural system is composed of subordinate faults derived from the Yanan Fault and the Luntai Fault, which have experienced multiple activities for long periods, with different activity intensities, properties, and oil control features [15, 16].

(1) The eastern section of this broom-shaped structural system is uplifted, while the western section is lowered; therefore, the eastern section is old and the western section is new in the distribution of the upper Pangeozoic and Proterozoic Systems, while the western section is old and the eastern section is new in the Mesozoic System. The Mesozoic strata directly cover the Paleozoic System, which is conducive to the formation of oil and gas reservoirs in the buried hills, including the Yakla Oil and Gas Field. The anticlinal or fault anticlinal structural zones formed along the Mesozoic faults are favorable places for oil and gas accumulation.

(2) Hydrocarbon source rocks are abundant, including the Cambrian–Ordovician hydrocarbon source rocks in the southern Aman Depression and the Mesozoic hydrocarbon source rocks in the northern Kuqa Depression. After the first or second periods of hydrocarbon generation in the Himalayan Period, hydrocarbon source rocks were migrated along the regional unconformity toward the structural zone and then migrated vertically in the event of faults and accumulated in various traps.

(3) More than 10 discovered oil and gas fields are distributed along the faults or the anticlines between two faults, showing a broom shape (spreading to the west and converging to the east), including in Yaha, Donghetang, Dalaoba, Yakla, Qiuli, the Sha 54# Well, Luntai, and Tiergen (see Figure 1).

1.1.3. *Oil and gas control of the tazhong broom-shaped structural system*

The Tazhong broom-shaped structural system is located on the Katak Uplift in the middle of the Tarim Central Uplift Zone, spreading to the north and converging to the southeast (see Figure 2).

There are many fault structural zones in the Tazhong area. Under various factors, the fault activities were obviously different in the different fault zones or different sections of the same fault zone. Based on seismic structural analysis, the type and the segmentation of the fault structural zone are studied and the causes of the fault are analyzed in combination with the regional structural evolution to investigate the oil and gas accumulation conditions in the main structural zones, analyze the role of faults in the oil and gas accumulation process, and discuss the relationship between the broom-shaped structural system and oil and gas accumulation.

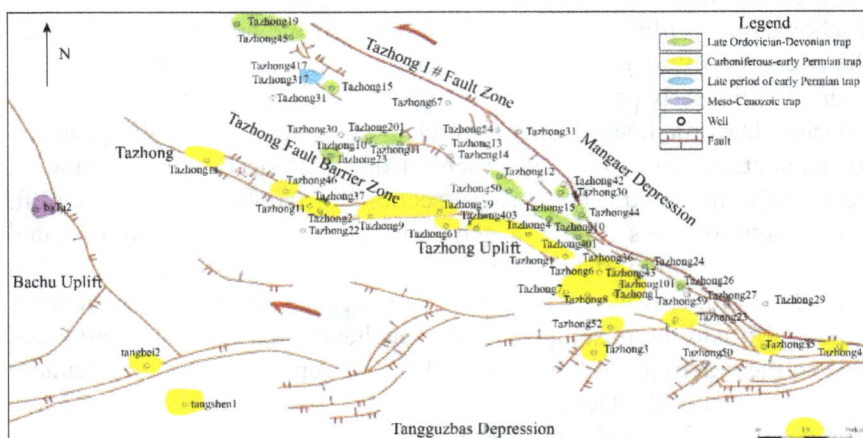

Figure 2. Layout of primary formation period of the main structural traps in the Carboniferous in the Tazhong area.

1.1.4. *Formation of the Tazhong broom-shaped structural system*

Based on the study of regional profiles, the stretching environment was transformed into a compressive environment under regional compression stress in the Tazhong area in the late Ordovician, resulting in the reversal of structural attributes and generating the Tazhong I# and Tazhong III# Faults, the Tangbei Fault System, and the Tumuxiuke Fault, among which the Tazhong II# Fault Zone had a large-scale extrusion-derived fault and the transition zone attributes. Therefore, the E-W-trending and N-W-trending paleo-faults of the Tazhong Uplift showed their original profiles [17, 18].

After large-scale stretching movement in the early Caledonian, the crust in this region changed from stretching to compression conditions at the end of the Middle Ordovician; during this period, the structural pattern of the Tarim Platform was E-W trending, which was framed by the Tazhong E-W-trending uplift zone. Due to the non-uniformity of N-S-trending compression stress, the western section of the Tazhong Uplift became wider, and the axis of the low uplift was spreading to the west in a finger-like manner. The early shear structure associated with the Tazhong I# Fault in the north and the retrograde diapiric fault zone and its associated retrograde dorsal slopes, which opened westward from the Tazhong III# Fault in the south, were formed at the same time. The Silurian and Devonian sediments were superimposed on an ancient erosion background, and the southern and northern depressions of the Tazhong Low Uplift were the earliest ones. At the end of the Devonian, the overall uplift of Tatong area further strengthened the structural pattern formed in the early period and increased the range of paleo-uplift, and the Silurian, Devonian, and Upper Ordovician Systems were again subjected to denudation. The strongest structural deformation and transformation occurred in the central and eastern sections of the Tazhong Low Uplift. The structural zones formed in the early stage were further complicated by the strengthening of faults, and the overthrust, uplift, and denudation were finalized. However, the southern and northern wings of the Tazhong Low Uplift and the nose uplift of the Gucheng Ruins were less affected. Due to the reactivity of the Tazhong II# Fault and the uplift and denudation at the end of the Devonian, the paleotopography of the Tazhong area was basically flattened, and the broom-shaped structural pattern of the buried hill was basically formed in the early Carboniferous deposition period. After the Carboniferous, the Tazhong area was basically in a stable state, maintaining a rise and fall on the whole, and only some faults to the center and west of the uplift were active. Due to the influence of late Hercynian Movement, passive emplacement were generated along the main fault of the broom-shaped structure. In the Himalayan period, due to the influence of the Altun left shear stress field, the Tazhong broom-shaped structure was finalized.

The regional structure controls the formation and evolution of fault structures in the Tazhong area [19]. On a plane, the fault structure shows a broom-shaped distribution in the Tazhong area. There is a large compound anticline and its two wings form the back thrust fault systems, forming a giant flower-shaped fault structure. The type of combination of each fracture tectonic zone includes positive "flower"-shaped, "Y"-shaped backwash, super-positional backwash, and counterstrike. The backbone fracture is basically a base-involved shape, and most of the faults are in a compressive shear state. The faults were mostly formed in the Caledonian Period and finalized in the Pre-Carboniferous Period. The active faults were mainly concentrated to the west of the uplift in the late Hercynian Period, and were basically finalized after the Permian Period. These features of the Tazhong broom-shaped structure were formed during the evolution of the regional geological structure (see Figure 2).

1.1.5. *Oil control of the broom-shaped structural system*

The oil control of broom-shaped structural system is mainly manifested in the restricting effect of the broom-shaped structure fault zone on the oil and gas reservoirs. The three groups of fault zones in the broom-shaped structural system have obvious control over the features of the paleo-topography, geostratigraphy, sedimentary facies, paleokarst, and reservoir caps, and over the migration and accumulation of oil and gas. The analysis of the oil and gas accumulation process shows that different fault structural zones have different controlling effects on oil and gas accumulation.

(1) **Tazhong I# Fault Structural Zone:** The slope-fold landform controlled by the Tazhong I# Fault in the middle and late Ordovician, that is, the platform margin, lays the foundation for the generation of a high-quality reservoir of the middle and upper Ordovician System, determines the near migration path of the zone, and then controls the oil and gas

accumulation horizon of the zone. At the same time, the karstification on the platform margin and the hydrothermal dissolution along the fault are conducive to the formation of pore and cavity reservoir spaces. For example, in the Tazhong 44# and 82# Wells, the accumulation of the Ordovician carbonate condensate gas is mainly controlled by reef and beach reservoirs. The Tazhong 45# and 30# Wells located on the derived fault of the Tazhong I# Fault are controlled by paleokarst and were formed by the leaching of atmospheric fresh water over multiple periods.

The activity of the Tazhong I# Fault was weak in the later stages, which is a guarantee of oil and gas accumulation. The fault basically had no strong activity after the Devonian Deposition Period, and there are several sets of high-quality regional caps present that are well developed, which play an important role in oil and gas accumulation.

(2) **Tazhong II# Fault Structural Zone:** Fracture and unconformity controlled by multiple differential lifts of fractures (and thus karst uplands) and reservoir modification by high-density fracture systems determine the development of high-quality pore-fracture-type reservoir space in the Middle and Lower Ordovician of the belt, which is one of the conditions for high hydrocarbon production in the Middle and Lower Ordovician.

The adjustment and the transformation of oil and gas reservoirs in multi-stage fault activities were important reasons for the redistribution of oil and gas in the early stage. The oil and gas reservoirs were re-injected in the Himalayan Period. On the one hand, the natural gas from the middle and lower Cambrian hydrocarbon source rocks and the oil and gas from the middle and upper Ordovician hydrocarbon source rocks were injected; on the other hand, due to fault activities, the C_{III} reservoir re-migrated along the fault to form C_I and C_{II} reservoirs with gas caps on the upper side. For Cambrian hydrocarbon source rocks, early oil gathering, late oil gathering, gas gathering, and adjustment and modification are essential features of this oil area.

The late fault activities provided a channel for vertical migration of oil and gas, which was key to the accumulation of oil and gas in the over-burdened system of a buried hill, and was also an important condition for the formation of complex oil and gas accumulation in this zone.

(3) **Tazhong 10# Well Fault Structural Zone:** In the Tazhong 10# Well structural zone, the fault was an important factor for the formation of a vertical migration channel and multi-source and multi-charge accumulation of oil and gas.

For oil and gas control in the main stage of oil and gas generation, the fault was active to open and connect the oil sources and the traps, thereby controlling the migration mode of oil and gas, which was quite common, as represented by the Tazhong 10# Structural Zone and the Tazhong Main Barrier Zone. For oil and gas control after the accumulation period, the fault became re-activated to control the adjustment of oil and gas reservoirs and the redistribution of oil and gas.

1.1.6. *Oil and gas control of the broom-shaped structural system*

At present, the industrial oil and gas discoveries in the Tazhong oil and gas area are mainly concentrated in the Tazhong I# Structural Zone, the east of the Tazhong II# Fault Structural Zone, and the northern slope of Tazhong (Tazhong 10# Well Structural Zone). The oil and gas sources are mainly concentrated in the Ordovician carbonate rocks and Silurian–Devonian clastic rocks. In October 1989, in a test on a 3,566–3,650-m section of the weathering crust on the top of the Ordovician System in the Tazhong 1# Well, a high-yield industrial oil and gas flow was discovered with a daily oil production of 356 m^2 and a daily gas production of 55.7×10^4m^3. By the end of 2013, eight oil and gas fields with proven reserves of 4×10^4t were discovered in the Katak Uplift Zone (Tazhong).

2. Oil and Gas Control of the Shear Structural System

2.1. *Oil and Gas Control of the Akkule Shear Structural System*

This shear structure is located in the Akkule Uplift in the middle section of the Shaya Uplift in the northern part of the Tarim Basin [20, 21]. It was formed in the Hercynian Period, consisting of two cycles and vortices. It plays an important role in controlling the distribution of oil and gas fields in this area (see Figure 3).

2.1.1. *Formation of the shear structural system*

(1) **In the Sinian–middle Ordovician,** a set of shallow marine carbonate rocks was deposited in the craton basin, and the structural activity was weak in this area. At the end of the middle Ordovician, N-S-trending

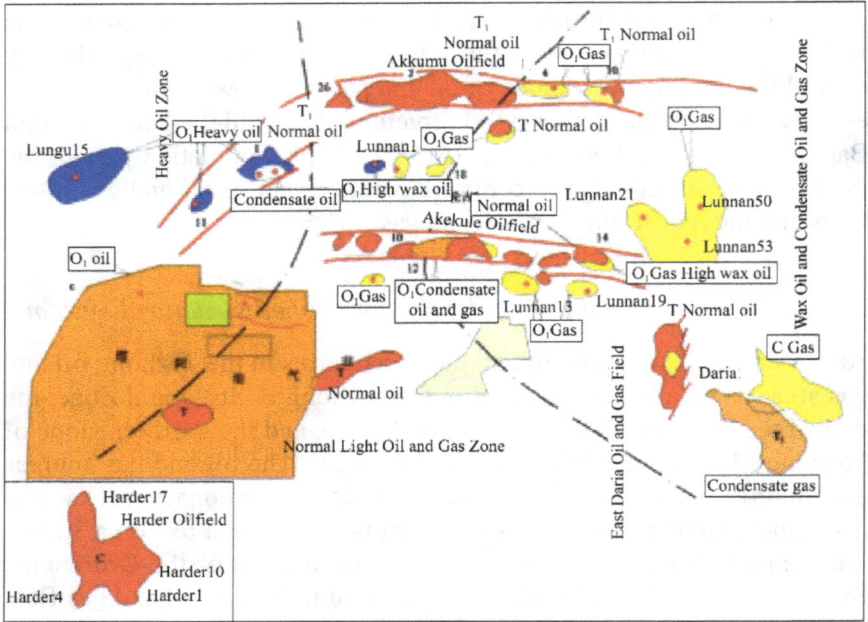

Figure 3.   Distribution of the Akekule shear structural system and the Tahe Oil Field in the north Tarim Basin.

compression occurred, resulting in uplift in this area and forming the Akku Small Fault Zone and the Sangtamu Fault Zone.

(2) **In the late Ordovician–Silurian,** the Tarim platform was in a compressive environment. During the development of the flexural basin, the Akekule Anticlinal Bulge was formed, the Akekumu and Santamu Faults were uplifted again, and the area was subjected to strong erosion.

(3) **In the Devonian,** under the continuous action of regional compression stress, the area was uplifted again, the fault activity intensified, and two back thrust fault structural zones were formed at the Akkumu and Akkule Faults.

(4) **In the Carboniferous–Permian,** the area began to receive sedimentation with platform extension and subsidence in the early and middle Carboniferous, forming a set of shallow marine carbonate rocks. The area was uplifted again in the late Carboniferous. Some volcanic eruptions also occurred in this area, accumulating matter in a set of pyroclastic strata, which were uplifted again in the late Permian and

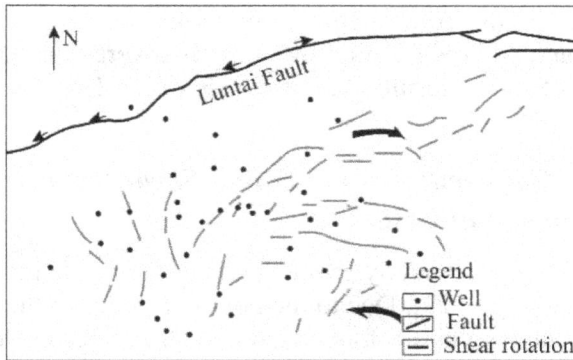

Figure 4.   Fault system in the Akekule Uplift in the late Hercynian.

subjected to intense denudation. During this period, the Akkule shear structure was basically finalized (see Figures 3 and 4).

(5) **In the Indosinian–Yanshanian,** structural activities were very weak in this area, and the Triassic System covered the strata formed in various periods. The Akkumu and Santamu Faults were still active and partially cut through the Triassic System.

(6) **In the Himalayan,** this area received deposits from the Cenozoic along with subsidence of the basin and was transformed into a slope dipping to the north.

2.1.2. *Oil and gas control of the structural system*

This tectonic system controls the distribution of hydrocarbons, mainly in (1) Lunnan oil and gas fields within the Outer cyclonic tectonic belt-Akkumu tectonic belt; (2) the distribution of Sangtamu oil and gas, and Dalimu oil and gas fields, etc., within the Inner cyclonic tectonic belt-Sangtamu tectonic belt; (3) The distribution of the Tahe Oil Fields in the swirling hydrocarbon aggregation area (see Figure 3).

Hydrocarbon source rocks of the abovementioned oil and gas fields are all from the Cambrian and Ordovician Systems, which were formed by long-term generation and expulsion of oil and gas, the migration and accumulation of the regional unconformities, and shear faults of the hydrocarbon source rocks. Oil and gas are distributed in reservoirs from the Ordovician, Devonian, Carboniferous, and Triassic.

The Tahe Oil Field is located in the inner cycle of the Akkule Shear Structural System of the Shaya Uplift. In October 1990, the S23# Well was designed in the Aixieke Structure (Tahe 3# Area) of the inner cycle zone, and high-yield oil flow was discovered in the Paleozoic Carboniferous System, resulting in discovery of the Tahe Oil Field.

2.2. *Oil and Gas Control of the Zhouji Shear Structural System in the Jianghan Basin*

The Zhouji Shear Structural System was mainly formed in the Neocaysian Structural System [22, 23]. Due to the non-uniformity of the stress field, the shear structural style was formed, consisting of three cycle zones (see Figure 5).

The inner and middle cycle structural zones adjacent to the center of the cycle structure had better geological conditions than the outer cycle structural zones, resulting in different oil-bearing conditions, and it is inferred that it should have some diagenetic connection with the action of the shear stress field associated with this shear tectonic genesis [24, 25].

The shear stress in this structural system was mainly caused by the uneven development of the partial shear stress activities in the primary structural plane of the new Cathaysian System. Since the shear stress is initiated from the periphery, and belongs to the passive knob, the structure of the system on each point of the intensity of the stress effect generally should be by the direction of the outer spinning layer to the inward weakening of the role of the moment is also getting smaller and smaller. This

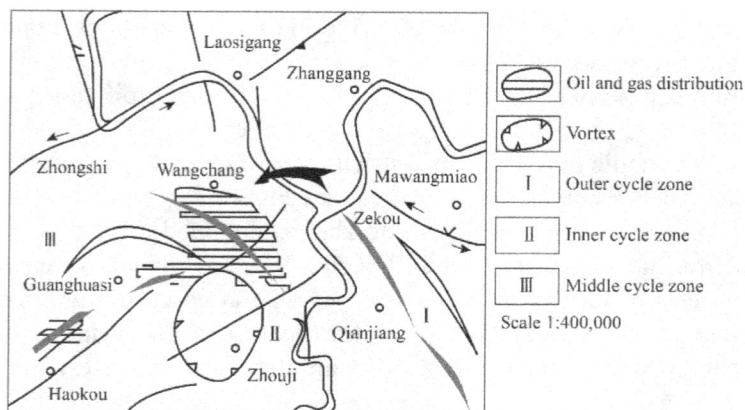

Figure 5.   Oil control of the Zhouji Shear Structural System.

shear stress accelerated the settlement velocity and amplitude of the syncline between the vortex and the structure zone, and improved the oil-
generation capacity in the original deposition center. At the same time, the
dispersed oil and gas were driven to the weak stress sites, and gradually
migrated and accumulated in the inner and middle cycle structural zones.
Because the component of the shear stress has an obvious effect along the
tangential direction of the cycle layer, and its strength also gradually
decreases inward from the outer cycle layer, and the even to zero near the
vortex site, where the current distribution of the oil and gas is found.

The formation and development period of the shear structure is also
the period of migration and accumulation of oil and gas. Since the third
portion of the Qianjiang Formation was deposited in the early Tertiary,
shear structure was formed and developed, and oil and gas also migrated
and accumulated in different layers of the structure, so the distribution
was strictly controlled by the structural system. Although there are appar-
ent changes in the oil and gas distribution in the second and third oil
groups of the fourth section of the Qianjiang Formation and the following
layers, for example, the third oil group of the fourth section of the
Qianjiang Formation is distributed contiguously in the tectonics of
Wangchang and Guanghuasi, the oil and gas enrichment is still mainly
around this tectonics and matches with the paleotectonics of the period of
the second section of the Qianjiang Formation, and the oil and water
boundaries can be seen to the north-western part of the Formation are
under control by the tectonic phenomenon [26].

Based on the analysis, accumulation of oil and gas in the 4th section
in a previous period is closely related to the early development of the N-E-
E-trending structure of a higher order and grade, but oil control of the
early structural system was influenced by the later development of the
cycle structure. In other words, during the formation and development of
the shear structure, the shear stress controlled the migration and accumulation of oil and gas in its range of influence and also had a certain influence on the oil and gas controlled by the previous systems. Oil and gas are
mainly distributed in the first inner cycle zone (see Figure 5).

3. Oil and Gas Control of the Echelon-shaped Structural System

A typical example of an echelon-shaped structural system is the Yecheng
echelon-shaped structural system in the southwest depression of the Tarim

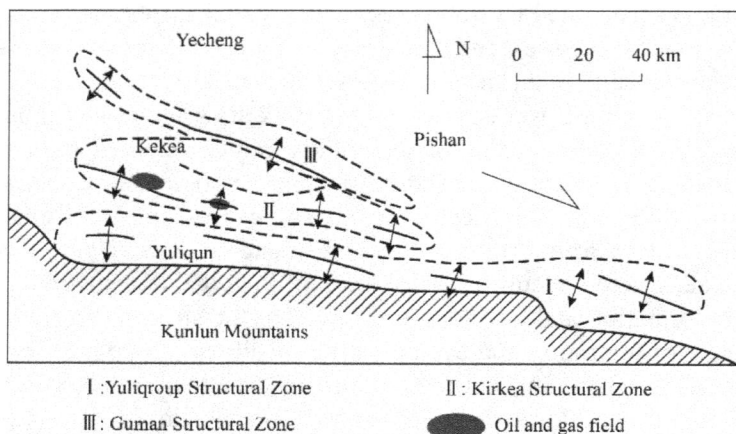

Figure 6.   The echelon-shaped structure and oil and gas field distribution in the Yecheng area.

Basin. The system is located in the Yecheng Depression and was formed in the Himalayan Period. It is an echelon-shaped structural system formed under intrabinal compression and shear of the Kunlun orogenic zone and consists of three structures: the Gumanian Structure, the Kekea Structure, and the Yuliqun Structure [27]. Exploration showed that the Kekea Oil and Gas Field was discovered in the Kekea structural zone with moderate ground stress and the Kedong Cretaceous Oil and Gas Field was recently found in the southeastern part of the second structure of the Kekea Oil and Gas Field, but no oil and gas fields were found in the southern and northern sections (see Figure 6).

4. Oil and Gas Control of the λ-shaped Structural System

4.1. *Oil and Gas Control of the Bashito λ-shaped Structural System*

The Bashito λ-shaped Structural System is located in the northern part of the Maigeti Slope of the Tarim Basin and consists of the N-W-trending

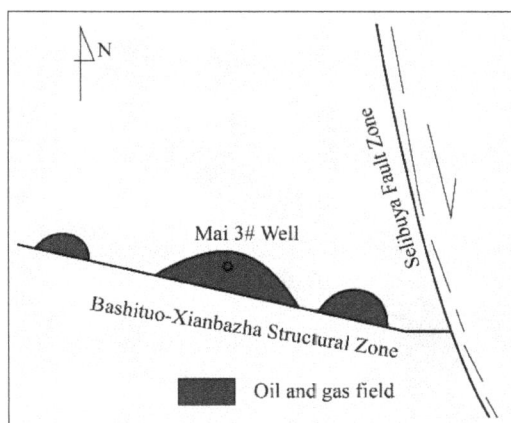

Figure 7.   The Bashituo–Xianbazha λ-shaped structure on the Maigaiti Slope.

Selibuya Fault Zone and the Bashito Structural Zone. This structural system was discovered by a 2D earthquake survey deployed by the Northwest Petroleum Bureau in 1991. It was formed in the late Hercynian Period and was identified as a shear structure. In the Mai 3# Well that was drilled in 1992, a major breakthrough in the oil–gas orientation was achieved in the top portion of the Carboniferous System when the Bashito Oil and Gas Field was discovered. Later, new discoveries were made along the structural zone to the northwest and southeast, expanding the oil-bearing area (see Figure 7).

The Bashito Oil and Gas Field was formed in the Himalayan Period, and the oil and gas originated from the Carboniferous–Permian System, mixed with a large number of Paleozoic Cambrian–Ordovician components, which makes it a typical λ-shaped structural system [28].

4.2. *Oil and Gas Control of the Beidagang λ-shaped Structural System in the Bohai Bay Basin*

The λ-shaped structural system consists of the main fault (the N-E-trending Taipingzhuang–Baishuitou Fault) and some branch faults (Liujianfang, Maxi, Madong, and Tangjiahe Faults) formed under left

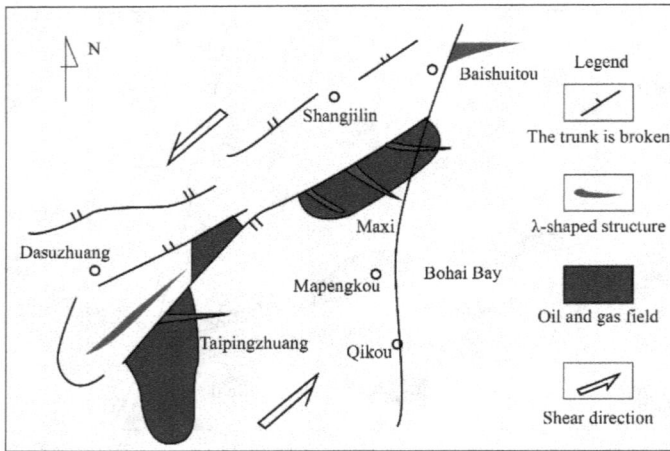

Figure 8.   The λ-shaped structure and oil and gas field distribution in Beidagang.

shear stress. The oil and gas fields are mainly distributed in the inner side of the λ-shaped structures (see Figure 8).

4.3. *Oil Control of the Renqiu λ-shaped and Echelon-shaped Structural System*

The Renqiu λ-shaped and echelon-shaped structural system is located in the Jizhong Depression in North China [29]. It is a long and narrow N-E-trending (30°) buried hill zone, the N-E-trending main fault has a 3,000-m altitude drop on the west side, and the east buried hill zone consists of 4 N-E-trending hilltops, which are controlled by branch faults intersecting with the main fault, forming an echelon shape. The main body of the buried hill is composed of siliceous dolomite of the middle and late Proterozoic Wumishan Formation. A Cambrian–Ordovician system is distributed on the northeastern hilltops with a monocline stratum, which is overlaid by the lower Tertiary system. Oil generated from hydrocarbon source rocks of the Shahejie Formation from the lower Tertiary system entered the trap along unconformity to form a buried hill oil field (see Figure 9).

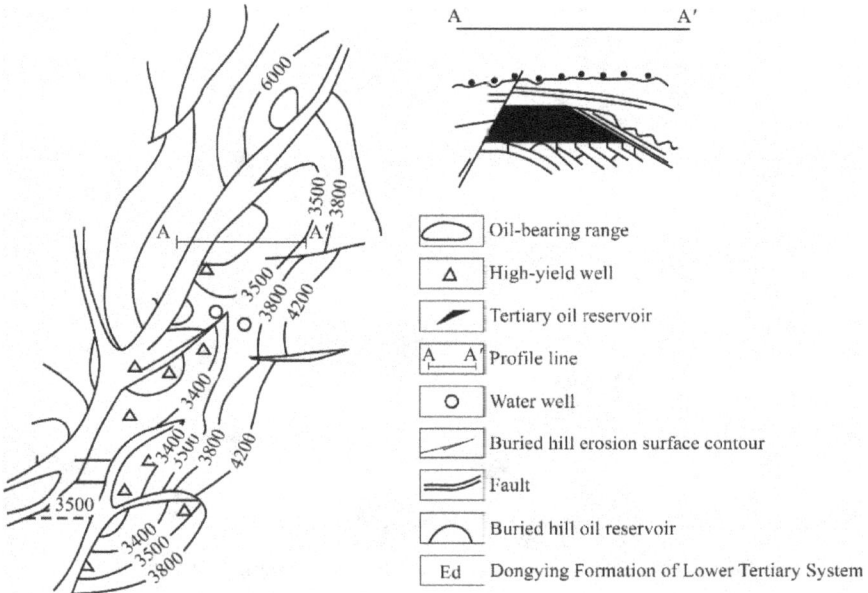

Figure 9. λ-shaped structure and cross section of the Renqiu Oil Field in the Jizhong Depression.

5. Oil and Gas Control of the Inverted-S-shaped Structural System

5.1. *The Hetianhe Gas Field in the Mazatak Inverted-S-shaped Fault Zone*

The Mazatak inverted-S-shaped structural system was formed in the late Hercynian Period and was active again and finalized in the Himalayan Period. In the Himalayan Period, a faulted anticline was formed and sandwiched by two faults in the north and south. The faults on both sides were active in the Himalayan Period, the main fault cut through the Miocene System, and the minor detachment fault on the southern side went straight to the sky along the Miocene and Pliocene Systems.

In the Himalayan Period, the natural gas generated from the Cambrian–Ordovician hydrocarbon source rocks migrated along faults upward from the depth into the middle Ordovician limestone and overlaid

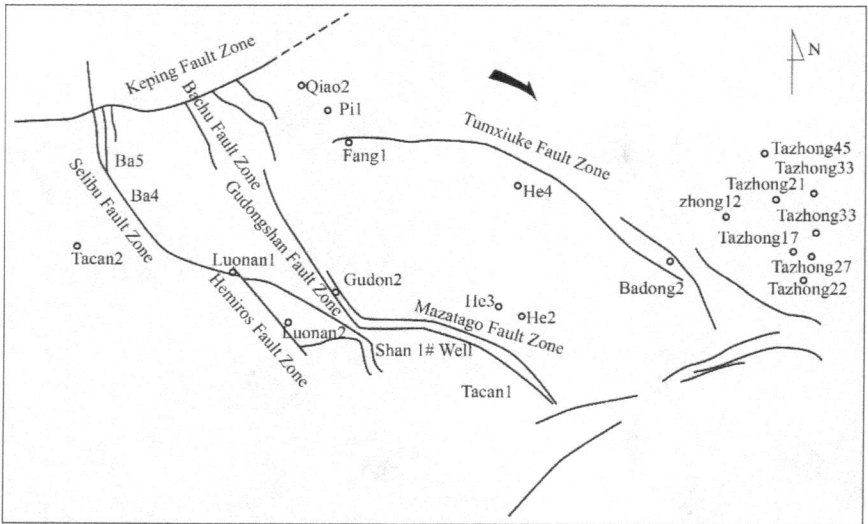

Figure 10.   The Mazhataka inverted-S-shaped (similar) structure.

Carboniferous limestone and sandstone, forming oil and gas reservoirs (see Figure 10).

(1) In October 1985, the Northwest Petroleum Bureau discovered the Yasongdi Structure in the northeast of the Libuya Fault Zone and drilled the Bacan 1# Well, which achieved the first major break-through in oil and gas orientation in the Bachu Uplift in the Carboniferous System and discovered the Yasongdi Oil and Gas Field.

(2) In February 1995, the CNPC drilled the Niaoshan 1# Well in the Niaoshan Structure of this structural system and obtained industrial gas flow from the lower Ordovician dolomite at a depth of 3,872–3,885 m, with an initial daily gas production of 4.1×10^4 m^3. In September 1997, the CNPC obtained industrial gas flow from the lower Ordovician limestone at a depth of 2017–2281 m in the Ma 4# Well on the Mazatak Fault Zone, with an initial daily gas production of 27.4×10^4 m^3. In 1999, the Ma 4# Gas Field was renamed the Hetianhe Gas Field, with proven gas reserves of 620×10^8 m^3.

5.2. *Oil and Gas Control of the Lenghu Inverted-S-shaped Structural System in the Qaidam Basin*

The Lenghu inverted-s-shaped system is located in the Structural Uplift between the Saishiteng and Kunteyi Depressions in the Sai-Kun Fault Depression in the northern margin of the Qaidam Basin and consists of the Lenghu 0–7# anticlines, and the Nanbace, Beilingqiu, Donglingqiu, and Nanjixing anticlines of different magnitudes [30–32]. These anticlines were obliquely echelon-shaped and inverted-s-shaped fold zones. There are N-N-W-trending faults to the north of the uplift zone axis, N-W-W-trending faults in the middle, and the Wuchaigou Fault in the northern end of the southern section which is in line with the hidden faults in the northern margin of the zone, but there is the N-N-W-trending fault cutting through the Nanjixing Fault to the west of the Wuchaigou Fault. The general fold axis is nearly EW-NW-NNW-NWW trending, reflecting the combined control of the compression southward of the faults in the northern margin of the Qaidam Basin and the overthrust strike-slip of the Altyn Fault. It is noted that there is an N-W-W-trending fault at the north end of the fold zone, which is one of the λ-shaped branch faults of the Altyn Fault. This N-W-W-trending fault intersects the Lenghu 0–2# anticline axis at an acute angle, reflecting its left strike-slip feature. This fault form the head section of the Lenghu Inverted-S-Shaped Fold Zone (see Figure 11).

Lenghu 7# is a simple anticline accompanied by faults in the same direction. As a whole, it is a fault anticline sandwiched between two faults, with a typical structural feature of "One Uplift Sandwiched Between Two Faults."

There is a set of basement faults of the same trend in the northern wing of the Lenghu inverted-s-shaped Structural Zone, with an intermittent extending length of 120 km. This set of basement faults has a great influence on the morphology and integrity of the anticline, resulting in a steep slope in the northern wing that is greatly damaged and a slow slope that might even be inverted on the southern wing. It is noted that oil immersion, oil mud, ground wax, asphalt, and other oil and gas components commonly exist along the fault. In particular, there is a large oil immersion range and raw oil was found at a lesser depth in anticline 4#, indicating a good oil storage structure. Oil fields have been found in Lenghu 3#, 4#, and 5# in this structural zone.

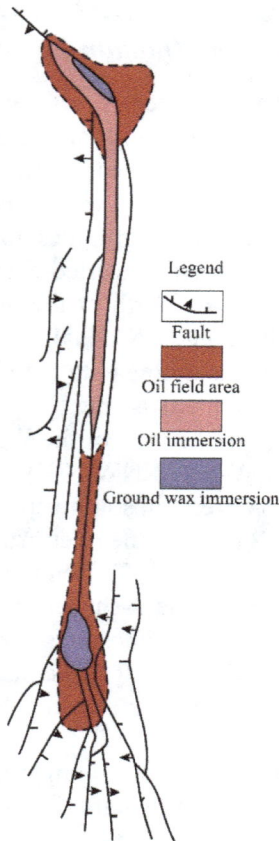

Figure 11.　Oil field distribution controlled by the Lenghu Inverted-S-Shaped Structural System in the Qaidam Basin.

In the Lenghu inverted-S-shaped Structural Zone, the small 0#, 1#, and 2# anticline structures were formed in late Pliocene–Quaternary, and other anticline structures were controlled by basement uplifts. They have the same formation features, the formation period of the northern section is dated before that of the southern section, the uplift amplitude of the northern section is higher than that of the southern section, the size of the northern section is smaller than that of the southern section, and both sections are closed in the head section and open in the tail section.

The oil fields are mainly distributed at the turns of the inverted-S-shaped structural system, which are the locations of the shear structure with moderate ground stress (see Figure 11).

5.3. *Gas Control of the Shiyougou S-shaped Structural System in the Southeastern Part of the Sichuan Basin*

The Shiyougou S-shaped structural system is located on the branch anticline zone of the Huaying Mountain Fold Zone in the southeast part of the Sichuan Basin. The anticline was a Triassic–Jurassic one formed in the Indosinian–Yanshanian Period, and was a part of the Neocayxia Structural System [33, 34]. Due to the uneven stress between the northern and southern sections, an S-shaped structure was formed.

The main gas-producing stratum is sandstone of the Triassic System in the Jialingjiang Formation, covered by multi-layer salt as a good cap. The gas fields are mainly distributed in the southern and northern turns of the S-shaped anticline, which are also its two high points, because moderate ground stress at turns is conducive to gas accumulation (see Figure 12).

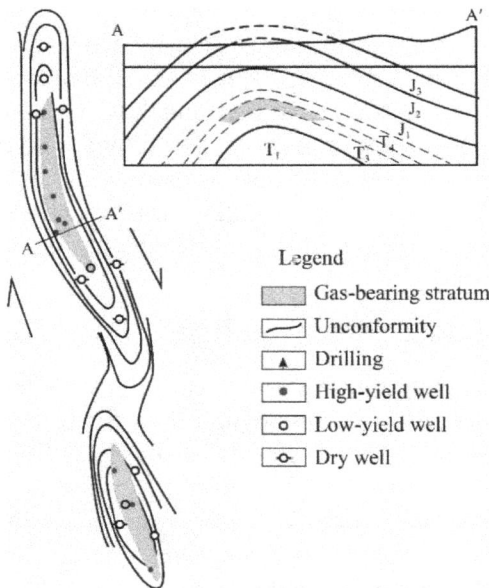

Figure 12.   Structure and profile of the Shiyougou–Dongxi Gas Field.

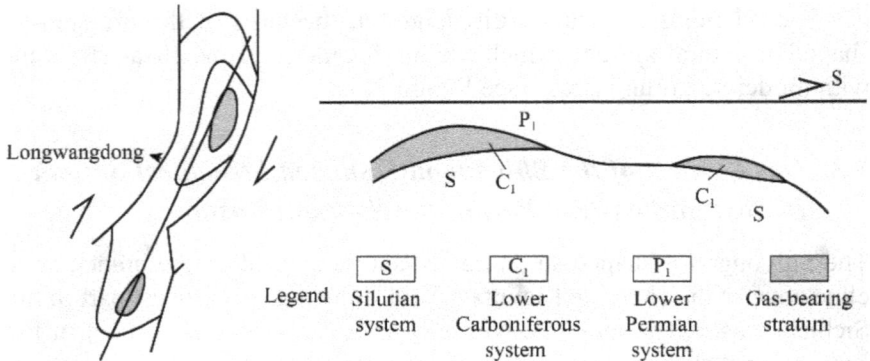

Figure 13.   Oil control of the inverted-S-shaped structure in the Xiangguosi Gas Field.

5.4. *Gas Control of the Xiangguosi Inverted-S-shaped Structural System in the Southeastern Part of the Sichuan Basin*

The Xiangguosi inverted-S-shaped system is located on the eastern side of the Huaying Mountain in the southeast part of the Sichuan Basin [35]. It was formed in the Hercynian Period and was still active and finalized in the Indosinian Period. The Xiangguosi Gas Field was formed in the turn of the inverted-S-shaped structural system in the Indosinian Period.

Oil was discovered and put into production in mid-1977. The reservoirs consist of upper Carboniferous dolomite and granulated limestone, with developed dissolution pores. Gas is mainly concentrated in the two high points of the anticline highs at the turns of the inverted-S-shaped structural system (see Figure 13).

6. Understanding and Conclusion

(1) Five kinds of subordinate shear structural systems are commonly found to control oil distribution, and their formation, evolution, and deformation features are controlled by the stress field during the formation of the corresponding structural systems.

(2) Different subordinate shear structural systems are formed under different structural environments and stress fields, which control the different distribution rules of the oil and gas fields.

(3) In a study of the structural systems, regional stress is analyzed, the change features of ground stress in different periods are clarified, and

the structural styles or types formed under the ground stress are studied to define the relationship between oil and gas accumulation conditions and structural systems, predict the favorable positions of oil and gas reservoirs, and identify favorable exploration directions.

(4) These five kinds of subordinate shear structural systems are widely distributed, and can be seen in all large basins; their control over oil and gas accumulation will be analyzed for future exploration of oil and gas.

References

[1] Siguang Li. *Introduction to Geomechanics.* Beijing: Science Press, 1973.

[2] Dianqing Sun, Naigong Deng. Prospect of oil resources in China as viewed from geomechanics. *Bulletin of the Chinese Academy of Geological Sciences*, 1979 1(1): 59–66.

[3] Dianqing Sun. Theory and practice of geomechanics in China petroleum survey. In *Bulletin of the Institute of Geomechanics CAGS (13)*. Beijing: Geological Publishing House, 1989.

[4] Yuzhu Kang. *The Relation between the Dominant Structural Systems and the Petroleum Distribution in China.* Urumqi: Xinjiang Science and Technology Press, 1999.

[5] Yuzhu Kang. Features of petroleum geology setting and oil and gas prospects of Northwest China. *Experimental Petroleum Geology*, 1984 6(3): 229–240.

[6] Shicong Guan. *Sedimentary Facies and Oil and Gas in Sea-land Changing Area in China.* Beijing: Science Press, 1984.

[7] Yuzhu Kang, Ximing Zhang, Zhihong Kang, *et al. Tarim Basin and Tahe Oil Field in China.* Urumqi: Xinjiang Science and Technology Press, 2004.

[8] Ci Bao, Kesheng Yang, Dengxiang Li. Features of geological structure and predication of gas prospect of Sichuan Basin. *Natural Gas Industry*, 1985 5(4): 1–11.

[9] Jianyi Hu. Geological foundation and oil and gas accumulation in Bohai Bay Basin. In *Collection of Mesozoic and Cenozoic Sedimentary Basins in China.* Beijing: Petroleum Industry Press, 1990.

[10] Yuzhu Kang, Zongxiu Wang, Xingui Zhou, *et al. Research on Oil Control of Tectonic System in Northwest China.* Beijing: China Land Publishing House, 2013.

[11] Yuzhu Kang, Hongjun Sun, Zhihong Kang, *et al. Paleozoic Marine Petroleum Geology in China.* Beijing: Geological Publishing House, 2011.

[12] Yuzhu Kang, Zongxiu Wang, Zhihong Kang, *et al. Study on Oil Control of Structural System in Junggar-Tuha Basin.* Beijing: Geological Publishing House, 2011.

[13]   Yuzhu Kang. The relation between tectonic system and oil and gas in Tarim Basin. In *Bulletin of the Institute of Geomechanics CAGS (9)*. Beijing: Geological Publishing House, 1989.

[14]   Yuzhu Kang. On regularity and oil-searching direction of oil and gas in Tarim Basin. *Earth Sciences — Journal of China University of Geosciences*, 1991 16(4): 429–436.

[15]   Yuzhu Kang. The discovery of high-yielding oil flow in well Shashen 2 and the direction of oil search in the future. *Oil & Gas Geology*, 1985 6(S1): 45–46.

[16]   Yuzhu Kang, Zhijiang Kang. Progress of geomechanics in oil and gas exploration in Tarim Basin. *Journal of Geomechanics*, 1995 1(2): 1–10.

[17]   Zerong Liu, Xiaoling Wang. More on Ji-Lu Brush Structural system. *Journal of East China Petroleum Institute*, 1981 1(2): 1–14.

[18]   Yuzhu Kang, Youyuan Huang, Zhongxian Zhang, *et al. Paleozoic Marine Oil and Gas Fields in Tarim Basin*. Beijing: China University of Geosciences Press, 1992.

[19]   Yuzhu Kang. *Petroleum Geology and Resource Assessment in the Northwest China*. Urumqi: Xinjiang Science, Technology & Health Press, 1997.

[20]   Siguang Li. *Convolute Tectonics and Related Compound Problems about Geotectonic System in Northwest China*. Beijing: Science Press, 1955.

[21]   Yuzhu Kang. *Features of Oil Formation in Chinese Paleozoic Marine Facies*. Urumqi: Xinjiang Science, Technology & Health Press, 1995.

[22]   Jiqing Huang. *Research on Geotectonic Features of China*. Beijing: Geological Publishing House, 1984.

[23]   Shujing Li. Division and characterization of main structural systems in China. In *Collection of Research on China's Provincial Tectonic System (1)*. Beijing: Geological Publishing House, 1985.

[24]   Baojun Liu. *Crustal Evolution and Mineralization in South China*. Beijing: Science Press, 1992.

[25]   Baojun Liu. Evolution of the South China Basin. In *Bulletin of the Institute of Chengdu Geology and Mineral Resources (14)*. Beijing: Geological Publishing House, 1991.

[26]   Bo Liu. Study on the mountain-type tectonic system in South Margin of South China. *Journal of Chengdu Geology College*, 1982 1(2): 45–53.

[27]   Yuzhu Kang, Zongxiu Wang, Huijun Li, *et al. Research on Oil Control of Tectonic System in Tarim Basin*. Beijing: China Land Publishing House, 2009.

[28]   Yuzhu Kang, Zhenwei Gan, Zhihong Kang, *et al. Petroleum Distribution and Exploration Experience in Major Basins in China*. Urumqi: Xinjiang Science & Technology Press, 2004.

[29]   Liangde Liu. Discussion on Hebei-Shandong direction structural belt. In *Bulletin of the Institute of Geomechanics CAGS (9)*. Beijing: Geological Publishing House, 1989.

[30] Yuzhu Kang, Zongxiu Wang, Zhihong Kang, *et al. Study on Oil Control of Tectonic System in Qaidam Basin.* Beijing: Geological Publishing House, 2011.

[31] Siguang Li. *Methodology of Geomechanics.* Beijing: Science Press, 1976.

[32] Dongxu Li, Jiyuan Zhou. *Introduction to Geomechanics.* Beijing: Geological Publishing House, 1986.

[33] Zhengwu Guo. *The Formation and Development of Sichuan Basin.* Beijing: Geological Publishing House, 1996.

[34] Mapping Group in Geomechanics Research in CAGS. *Specification of the Structural System Map of the People's Republic of China (1:400,000).* Beijing: Geological Publishing House, 1978.

[35] Yuzhu Kang, Zhihong Kang, Zhijiang Kang, *et al. Introduction to the Global Tectonic System.* Beijing: China Petrochemical Press, 2018.

Tectonic Systems in Northwestern China and Their Relations with Hydrocarbon

(To celebrate the 30th anniversary of *Xinjiang Petroleum Geology*)

Abstract

Northwestern China refers to the area to the west of Helan Mountains and to the north of Kunlun Mountains, with an area of about 260×10^4 km^2. More than 60 sedimentary basins have been developed within this area, which is about 130×10^4 km^2. Up to now, more than 80 oil–gas fields have been discovered, of which 13 are giant oil–gas fields indicating that this area is very rich in petroleum resources with an equivalent of 440×10^8 t oil; but, this area also has an extremely low resource transformation ratio (about 8%), showing that it has tremendous potential for petroleum exploration and could be one of major explorative areas for replacement of resources. In view of the structural system, the evolution of petroliferous basins, and control factors of oil–gas distribution, this chapter discusses the oil–gas distribution rules and indicates the directions or targets for petroleum exploration in this area in the future.

Keywords: China, northwest, petroliferous basin, tectonic system, and petroleum distribution.

Under the guidance of Li Siguang's geomechanics theory [1], this chapter discusses the distribution, development, and evolution of tectonic systems in Northwestern China and their relations with the rules of oil and gas distribution based on the idea of oil and gas exploration by tectonic systems, with the aim of exploring this area in order to find replacements of resources in China as soon as possible.

1. Tectonic Systems and Their Characteristics

1.1. *Division of Tectonic Systems*

The main tectonic systems in northwestern China are divided into the latitudinal tectonic system, the Western Region tectonic system, the Altun Mountain tectonic system, the Pamir inverse-S tectonic system, the new Western Region tectonic system, the quasi-WNE tectonic system, and the Helan Mountain ε-type tectonic system. These tectonic systems have evolved to control the formation and development of the petroliferous basins in this area [2].

1.1.1. *Kunlun Mountain latitudinal tectonic system*

This tectonic system, which is part of the Kunlun Qinling latitudinal tectonic system, is mainly distributed in the Kunlun Mountains on the southern edge of the Tarim Basin, the Kumukuli Basin, and the Qaidam Basin. Due to the influence of the η-type structure of Pamir and Qinghai–Xizang–Sichuan–Yunnan province, the tectonic system waves and bends, and moves northward from the normal position, roughly distributed at 34°~36°30′N. The Kunlun Mountain latitudinal zone is cut by the Altun fault zone into two parts, i.e., the East Kunlun Mountain latitudinal zone and the West Kunlun Mountains latitudinal zone.

(1) **The East Kunlun Mountains latitudinal zone,** located in the central part of Qinghai, consists of a series of east–west mountain systems, such as the Hoh Xil Mountains, the A'erge Mountains, the Burhan Budai Mountains, and the Xiqing Mountains. This zone stretches for more than 1,000 km and is about 200 km wide. This zone has grown and developed after multiple violent tectonic movements. It is mainly composed of structural features such as compound folds and

compressive fractures with nearly east–west strikes, intermediate-acid intrusive rock belts, and masses from the Late Paleozoic and the Mesozoic, and can be divided into two sub-zones, namely, the northern sub-zone and the southern sub-zone. The northern sub-zone consists of a huge east–west rock belt composed of the metamorphic rock series of the Sinian Suberathem and the Lower Paleozoic, and intermediate-acid and acidic intrusive rocks of the Late Paleozoic and the Mesozoic. Its main body is a complex anticline involving the A'erge Mountains and the Burhan Budai Mountains. The southern sub-zone consists of rock series of the Mesozoic and the Upper Paleozoic, with extremely undeveloped igneous rocks. Its main structural features include the Hoh Xil syncline, the Kunlun Mountain pass, and the Xiqing Mountain anticline. Most fractures of this system occur in parallel bundles. Due to the interference of other systems, this system is not continuous in the longitudinal direction. The fracture surfaces generally dip to the north, and phenomena like nappe overthrust are commonly seen. On the main fault zone, there are beaded distributions of Late Paleozoic ultrabasic and basic rocks. Most of these major faults are characterized by southward or northward translations and dislocations. This system connects with the West Qinling Mountains to the east and extends south of Xinjiang to the east of the Pamirs.

The distribution and structural features of the Sinian System and the Lower Paleozoic indicate that the east–west zone may have taken shape in the Early Paleozoic. The Silurian System is overlaid discordantly by the Devonian and Carboniferous Systems in the area of the Burhan Budai Mountains. This implies that there was strong tectonic movement in the Late Caledonian Period, which formed the east–west structure and controlled the Late Paleozoic sedimentation. The tectonic changes and magmatic activities at the end of the Paleozoic Era still demonstrated the leading role of the east–west zone, and frequent and intense magmatic activities formed the northern east–west granite zone. The Triassic, Carboniferous, and Permian had discordant contact, indicating that the system matured in the Late Hercynian Movement. The tectonic movement at the end of the Triassic enabled the East Kunlun region to complete the sea and land changes. Throughout the Mesozoic to Early Cenozoic, several tectonic solid movements in the region have involved the middle and new Cenozoic boundaries around Hoh Xil in this latitudinal tectonic system and several continental volcanic eruptions.

(2) **The West Kunlun Mountain latitudinal zone** is mainly located between the Tiekelike northern edge fault and the Kangxiwar fault, reaching the edge of the Yecheng Depression in the north, and separated from the Karakoram Range in the south. It can be roughly divided into three sub-zones: the Tiekelike uplift zone in the north, the Gonger, Muztag, and Sangzhutag uplift zones in the middle, and the Hercynian trough fold zone in the south. This zone extends to Pamir in the west and is cut by the Altun Fault in the east. Its evolutionary history and structural characteristics are similar to those of the East Kunlun Mountain latitudinal zone.

1.1.2. *Tianshan Mountain latitudinal tectonic system*

The Tianshan Mountain latitudinal tectonic system is distributed in the Tianshan Mountains in the central part of Xinjiang, roughly at 40–44°N. Due to the the diagonal composite of the Borokonu Belt of the Western system which spreads in a north-westerly direction, the Tianshan Mountain latitudinal tectonic system not only shifts northward, but spreads in scope, reaching the southern edge of the Junggar Basin to the north and the northern edge of the Tarim Basin to the south. It mainly consists of the North Tianshan (the Yilianhabirga Mountain and the Bogda Mountain) anticline, the Middle Tianshan anticline, the South Tianshan anticline, and the Yining–Tuha syncline as well as their parallel fault zones. It is associated with well-developed torsional fractures in the NE and NW groups, sometimes forming obvious torsional fault zones. It should be pointed out that the Middle Tianshan anticline acts as a skeleton in the development of geological structures in the Tianshan area. The basement is mainly composed of a set of Precambrian metamorphic rock series. In the Cambrian–Silurian Period, the Tianshan area is mainly composed of normal clastic and carbonate facies deposits, occasionally mixed with volcanic tuffs. In the Silurian Period, the central part began to rise, dividing the Tianshan area into two different parts, i.e., the north Tianshan area and the south Tianshan area.

(1) **In the north Tianshan area,** clastic and volcaniclastic rocks were relatively developed in the Devonian Period. In the Carboniferous Period, volcanic rocks were dominant, while normal deposits were rare. In the Early Permian, clastic rocks of marine and continental facies mixed with volcanic tuffs were formed, and in the Late Permian, there were deposits of continental facies.

(2) **In the south Tianshan area,** the Devonian mainly saw clastic rocks and carbonate sediments, the Carboniferous was characterized by littoral–shallow marine clastic rocks with gypsum and bauxite being generated, and the Permian was characterized by marine volcanic rocks. The corresponding acidic and basic rocks were widely developed and distributed in the Tianshan area. Regional metamorphism and migmatization are also obvious, closely related to the faults and their direction of distribution is consistent with the tectonic line. As for the whole Middle Tianshan area, from the Silurian to the Permian, except for a small amount of iron–manganese-bearing glutenite and limestone deposits in the Early Carboniferous, other areas in Middle Tianshan were denuded. The Carboniferous System was in discordant contact with both the upper and lower strata.

To sum up, the Tianshan Mountain latitudinal tectonic system features complex, long-lasting formation and development. During the Sinian Period, the sedimentation was obviously controlled by the east–west trough, indicating that the Tianshan Mountain latitudinal tectonic system had taken shape at that time. In the Paleozoic Era, this control became increasingly obvious, as evidenced by the distribution of Cambrian phosphate rocks. The strong deformation, magmatic activity, and metamorphic migmatization in the Late Paleozoic were some of the important features of this system, and magma intrusion and eruption were extremely strong. Acidic or basic rock type intrusions are generally distributed in the east–west direction.

This tectonic system controls the Central Uplift and Depression and the Southern Changji Depression of the Junggar Basin; the Yining and Tuha Basins in the Tianshan Mountains; and the Kuqa Depression, the Shaya Uplift, the Central Uplift, and the Yecheng Depression in the Tarim Basin.

1.1.3. *Western Region tectonic system*

Back in 1939, Li Siguang called the NW-NW tectonic system in western China the "Western Region-Xizang System" [2].

The NW and NWW tectonic lines included in the Western Region system are distributed in a strictly regional scope.

The Western Region system consists of a series of parallel and roughly equidistant NWW tectonic zones, each of which has a long history of development. They emerged in the Proterozoic, became active in the Early Paleozoic, took shape in the Late Devonian, and continued their activities in the Mesozoic and Cenozoic. They are dominated by uplift zones composed of large compression-torsional faults and complex folds with a strike of 270~310°, with a series of subsidence zones composed of basins that developed on both sides, which are arranged in a zigzag pattern. These NWW tectonic zones have experienced significant dextral torsion [3, 4].

There are three complex tectonic zones in Northwest China, namely, the East Junggar complex tectonic zone, the Boluohuoluo–Qilian complex tectonic zone, and the Bachu–Qimantag–Supilin complex tectonic zone, which control the evolution of the Junggar Basin, the Tarim Basin, the Qaidam Basin, and other basins in the corridor area.

In the vast area among these huge and complex tectonic zones, there is a NWW relative subsidence zone. The Junggar and Turpan Basins lie between the East Junggar and Boluohuoluo complex tectonic zones; the Tarim and Qaidam Basins lie between the Boluohuoluo complex tectonic zone and the Bachu–Qimantag–Supilin complex tectonic zone, where the Mesozoic and Cenozoic strata are deposited. The structures in the NW direction are relatively developed, accompanied by latitudinal tectonic systems and the Altun tectonic system, resulting in several basins that are either isolated or connected.

1.1.4. *Altun Mountain tectonic system*

The Altun Mountain tectonic system is composed of a series of large compressive and compression-torsional faults, linear folds, sedimentary troughs, strip intrusions, and eruptive rocks distributed in the northeast direction. It is particularly strong in the Altun and Beishan areas, with strong shear force and sinistral torsion. It appeared in the late Hercynian movement, and its NE-trending fault controlled the distribution of Carboniferous and Permian sediments and volcanic rocks. Its Mesozoic and Cenozoic activities were quite intense, and during the Himalayan Movement Period, it was further shaped. The system can be divided into three sub-tectonic zones, namely, the Altun Uplift, the Qiemo–Ruoqiang Fault, and the Beiminfeng–Luobuzhuang fault.

1.1.5. *New Western Region tectonic system*

The New Western Region tectonic system is generally distributed in the NNW direction. Originating from the Indosinian Movement and finalized in the Late Himalayan movement, it was composed of NNW fault zones and troughs, and characterized by composite interpenetration, transformation, and utilization of the pre-Mesozoic structural features.

1.1.6. *Pamir inverse-S tectonic system*

Located at the southwest edge of the Tarim Basin, the Pamir inverse-S tectonic system is composed of many fold zones and compressive and compression-torsional fault zones around the Pamirs and Himalayas. The southwest depression in the basin is a subsidence tectonic zone around the inverse-S system. It was part of the Western Region and latitudinal tectonic systems in the Paleozoic. Since the Mesozoic–Cenozoic Era, the Pamir inverse-S tectonic system has utilized and transformed the original tectonic system, so it is actually a multi-system composite. According to the tectonic systems and sedimentary characteristics, the structural units in the basin have been divided [4].

1.1.7. *Quasi-WNE tectonic system*

The Quasi-WNE tectonic system is mainly distributed in the northwest edge of the Junggar Basin. Trending in the northeast direction, it is a Eurasian ε-type east-wing reflection arc. It took shape during the middle and late Hercynian Movement. As a thrust belt composed of multiple thrust faults, it is about 300 km long and 20~30 km wide, and the horizontal nappe distance is estimated to be 25 km. It was still active in the Mesozoic and the Cenozoic, and this thrust fault zone has obvious control over oil and gas migration and accumulation.

1.1.8. *Qilian arc-type tectonic system*

This arc-type tectonic system is in the Helan Mountain ε-type reflection arc area, and is compounded with the Western Region tectonic system. It emerged in the Devonian period, initially took shape in the Carboniferous–Permian Period, and was finalized in the Cretaceous. It consists of a series of arc-shaped tectonic zones and faults, and controls the formation and development of Meso-Neozoic basins in the Hexi Corridor area.

1.2. Composition of Tectonic Systems

The latitudinal tectonic systems in Northwestern China began to appear in the Proterozoic Era or earlier, with the most intense activities in the Caledonian Movement Period, and during the Himalayan Movement Period, the systems became active again. These systems play an important role in the evolution of the Tarim block and basin.

The Western Region tectonic system emerged mainly during the Caledonian Movement Period, and its activities were the most intense during the Hercynian Movement Period, and lasted till the Himalayan Movement. This system is obviously equitangential to the latitudinal systems.

The Altun tectonic system was formed during the Caledonian Movement Period, and was strongly active during the Hercynian Movement Period. It was still obviously active during the Himalayan Movement Period. The system is oblique to the latitudinal systems.

The Pamir inverse-S tectonic system was formed during the Yanshan Movement Period, and was the most active during the Himalayan Movement Period. It was compounded with the latitudinal systems and the Western Region system.

1.3. Characteristics of the Tectonic Systems

1.3.1. Multiple phases of tectonic movement

There are six major regional unconformities between the Sinian Period and Pre-Sinian Period, Silurian and Ordovician, Carboniferous and Devonian, Triassic and Permian, Cretaceous and Jurassic, and Neoproterozoic and Palaeoproterozoic in the area. According to the geo-dynamic background, tectonic location, paleogeomorphology, and other factors, these unconformities can be divided into the following categories: (1) the near-source collision-type unconformities of blocks: high intensity and denudation thickness, usually distributed in large areas, like the clo-sures of the Kudi Ocean, the South Tianshan Ocean, the Paleotethys Ocean, and the Paleo-Continental Ocean in the region, with correspond-ing unconformities between the Carboniferous and Devonian, and the Triassic and Permian; (2) near-source discrete unconformities of blocks: relatively high tectonic intensity, mainly formed in the process of slow rises and falls, overlapping unconformities usually developed along the paleo-uplifts or slope areas (zones), and also distributed in large areas, like the unconformities between the Cambrian and Sinian, the Upper Ordovician and Middle Ordovician, and the Silurian and Ordovician; and (3) distant-source collision-type unconformities of blocks: low tectonic intensity, low denudation thickness, small spatial distribution, with some local distribution, like the unconformities between the Cretaceous and Jurassic, the Paleogene and Cretaceous, and the Neogene and Paleogene.

1.3.2. *Diversity of tectonic deformation*

As a result of the multi-phase complexity of the tectonic movement in the area, the tectonic styles of the basins are really diverse, which can be divided into the following categories: (1) compressive tectonic style, (2) shear tectonic style, (3) overlay tectonic style, and (4) wire tectonic style.

1.3.3. *Migration of tectonics*

(1) **Migration of tectonic intensity:** The tectonic movement in western China was strong in the north and weak in the south in the Paleozoic, and strong in the south and weak in the north in the Mesozoic and Cenozoic. The tectonic intensity in the Mesozoic and Cenozoic was strong to the west and weak to the east of the Tarim Basin, strong to the north and weak to the south of the Turpan–Hami Basin and Qaidam Basin, and strong to the south and weak to the north of the Junggar Basin and Corridor Basin.

(2) **Migration of sedimentary centers:** The sedimentary center of the Tarim Basin was in the Manjiaer area in the northeast of Tarim in the Early Paleozoic, and later migrated to the Yecheng area in the southwest of Tarim. The sedimentary center was in front of the Tianshan Mountains and in the center of Tarim in the Triassic, while in the Jurassic, it moved to the northeast of Tarim in addition to the piedmont area, and in the Neozoic, it was to the southwest of Tarim. The sedimentary center of the Junggar Basin was located in the Urho region on the northwestern edge and in the southern part of the basin, west of Urumci. In the Triassic, the sedimentary center moved to the north of Urumqi; in the Jurassic, it moved to the Changji–Manas region on the southern edge; in the Cretaceous, it moved to the center of the basin; and in the Paleogene, it moved to Manas–Wusu

region. In the Late Permian, the sedimentary center of the Tuha Basin was located to the north of Hami; in the Triassic, it moved south to the Hami area; and in the Jurassic, it moved to the northern depression. In the Jurassic, the sedimentary center of the Qaidam Basin was located on the northern edge; in the Paleogene, it was located in the southwest; and in the Neogene–Quaternary, it migrated to the east of the basin.

1.4. *Zoning of the Tectonics*

Northwestern China sees very clear tectonic zoning of superimposed basins. In the Paleozoic cratonic basin, this zoning is manifested as alternating arrangements of large uplifts and depressions, such as the Shaya Uplift, the Manjar Depression, the Central Uplift, and the Tanggula–Bayankala Mountain piedmont depression; and the Junggar Basin presents the Changji Depression, the Central Uplift, the Central Depression and the Luliang Uplift in order from south to north. The tectonic zoning in the foreland basin is also very clear. For example, the Kuqa Foreland Basin in the north of the Tarim Basin can be divided into an overthrust tectonic zone, a fault fold tectonic zone, a depression zone, and a slope zone from north to south; and the foreland basin on the southern edge of the Junggar Basin can be divided into an overthrust tectonic zone, a fault fold tectonic zone, a depression zone, and a slope zone from south to north.

1.4.1. *Cenozoic compression-torsion*

In Northwestern China, the Indosinian and Himalayan Movements, especially the latter, have shown significant compression and compression-torsion, impacting the tectonic deformation of the basins by strong folding, shortening, rapid uplifting, and severe subsidence of the orogenic belt, as well as formation of a fold-thrust belt. The cratons in the basin have also been remodeled and reshaped during the movements, resulting in new changes and formations in the tectonic framework and oil- and gas-bearing system across the basin.

1.4.2. *Variable basin–mountain compositions*

(1) The overthrust nappe-type is manifested as the deformation characteristics during the insertion of the land crust to the orogenic belt,

resulting in the uplift of the orogenic belt and pushing to the basin, Bogda tectonic zone and the northern edge of the Tuha Basin, the Qilian orogenic zone and the southern edge of the Jiuquan–Minle basins, the Kunlun orogenic zone and the southwestern edge of the Tarim Basin.

(2) The thrust and overlap alternating type is represented by the north-western edge of the Junggar Basin and the Kongquehe slope area of the Tarim Basin that feature alternating thrust and overlap from the Mesozoic and Cenozoic.

(3) The coupling contact of tectonic uplifts in different directions, pre-senting in the southwest depression of the Tarim Basin, the northwest end of the Bachu uplift, and the Awati fault depression, vertically contact the western Tianshan tectonic zone (including Keping uplift); and the western Qaidam Basin and the Altun tectonic zone also inter-sect nearly vertically.

(4) The nappe plus strike-slip type is represented by the composition between the Altun uplift and the southeast Tarim uplift. Other exam-ples include coupling between the Qaidam Basin and the Dunhuang Basin. Analysis of the basin coupling geostress in Northwestern China reveals that there are three different stress relationships: (1) compressive stress contact, such as the composite relationship of the Tianshan tectonic zone with the southern edge of the northern Junggar Basin and the northern edge of the Tarim Basin; (2) compres-sion torsional stress contact, such as the coupling between the south-west part of the Tarim Basin and the Kunlun Mountain tectonic zone; and (3) a basin–mountain complex dominated by torsional stress, such as the stress complex coupling of the Altun tectonic zone with the southeast edge of the Tarim Basin and the northwest edge of the Qaidam Basin [5].

2. Tectonic Systems Control Basin Evolution

As a result of a combination of multiple, multi-stage activities of tectonic systems, the oil- and gas-bearing basins in Northwestern China have evolved in multiple cycles, roughly in seven stages: Z-O tension rift–craton, S-D compressive rift–craton, C-P tension–craton, Mesozoic and Cenozoic foreland basin evolution stage, T-J foreland fault basin, K-E depression basin, and N-Q compression-torsion basin (Table 1).

Table 1.   Evolution of the major basins in Northwestern China.

Ages	Tarim Basin	Junggar Basin	Tuha Basin	Corridor Area	Qaidam Basin
Miocene (N₁) – Quaternary (Q)	Intracontinental unified basin	Atrophic basin	Intermountain basin	Intermountain basin	Atrophic basin
Cretaceous (K) – Paleogene (E)	Late depression basin	Late depression basin	Late depression basin	Late depression basin	Late depression basin
Triassic (T) – Jurassic (J)	Early fault basin	Early depression basin	Early fault basin	Early fault basin	Early fault basin
Carboniferous (C) – Permian (P)	Extensional cratonic depression basin	Extensional cratonic depression basin	Extensional cratonic depression basin	Extensional cratonic depression basin	Extensional cratonic depression basin
Silurian (S) – Devonian (D)	Compressional flexure basin	Compressional rift cratonic basin	Compressional flexure basin	Compressional flexure basin	Compressional flexure basin
Sinian (Z) – Ordovician (O)	Extensional rift cratonic basin	Extensional rift cratonic basin	Extensional rift cratonic basin	Extensional rift cratonic basin	Extensional rift cratonic basin

3. Oil Control Function of Tectonic Systems

3.1. *Tectonic Systems Control Source–reservoir–cap Combinations*

(1) **Multi-age source rocks:** Multi-system source rocks have developed in all major sedimentary basins in Northwestern China, but the Tarim Basin is the most developed with source rocks from the Sinian to the Neogene System (except the Devonian System) [3]. However, there are fewer hydrocarbon source beds in small and medium-sized basins.

According to preliminary calculations, the main basins in Northwestern China are rich in oil and gas resources, with an oil equivalent to $50 \times 108t$ oil [6].

(2) **Multi-age reservoirs:** The reservoirs in this area feature multiple ages and layers (Z, O, S, D, C, P, T, J, K1, E, and N1), and consist of four rock types: clastic rocks, carbonate rocks, volcanic rocks, and metamorphic rocks. Clastic and carbonate rocks are dominant, but oil and gas fields have been discovered in all the four types of reservoir rocks [7].

(3) The cap rocks in the area also feature multiple layers (Z-N1) and are of different ages. The main rock types are dense limestone, mudstone, marlstone, silty mudstone, gypsum, salt rock, volcanic lava, and dense sandstone.

(4) **Multiple sets of reservoir–cap rock combinations:** In the northwest region, five sets of source rocks (ε-O, C-P1, T3-J2, K2-E, and N1) have mainly developed from bottom to top, forming corresponding source–reservoir–cap combinations around each set of source rocks [3].

(5) **Multi-age reservoir forming:** The superposition of multiple types of basins results in multiple sets of hydrocarbon source rocks, multiple phases of hydrocarbon production, and multiple phases of reservoir formation. In the Tarim Basin, there are four reservoir-forming periods: the Late Caledonian Movement–Early Hercynian Movement, the Late Hercynian Movement, the Indosinian–Yanshan Movement, and the Himalayan Movement. In the Junggar Basin, there are at least two reservoir-forming periods: the Yanshan Movement and the Late Himalayan Movement. Other small and medium-sized basins mainly formed reservoirs during the Himalayan Movement.

3.2. *Tectonic Systems Control Oil and Gas Distribution*

(1) **Paleo-uplifts:** Examples include the Shaya Uplift and the Katalone Uplift in the Tarim Basin, as well as the Central Uplift, the Western Uplift, and the Luliang Uplift in the Junggar Basin. A total of more than 60 oil and gas fields have been discovered, of which 40 are distributed on the uplifts. The main reasons are as follows: (i) The uplifts are accompanied by large sedimentary depressions on both sides, and sedimentary depression facies of different ages and types are superimposed, forming large sedimentary depressions through long-term development. These depressions have developed multiple sets of source rocks of different ages to varying degrees, becoming important oil and gas source areas. The uplift is surrounded or sandwiched by the hydrocarbon source area, and it is located at the high altitude of the tectonic structure, which is the pointing area for hydrocarbon transport and enrichment. The area simultaneously receives oil and gas transported from its own source area and from source areas on both sides and accumulates these into reservoirs. Exploration of the Yakra–Luntai Faulted Uplift in the Shaya Uplift has confirmed that there is not only Paleozoic marine crude oil that migrated from the southern Awati–Mangar Depression but also Mesozoic continental oil and gas that migrated from the northern Kuqa Depression. (ii) During the geological development history, the uplifts, located at relatively high structural positions over a long period, were sometimes exposed on the surface and subjected to denudation, thus becoming the source areas of sediments in adjacent depressions; sometimes the uplifts sank underwater and received shallow-water coarse debris deposits. Therefore, not only do the uplifts and both their sides develop good reservoir rocks but the physical conditions of the reservoirs are also greatly improved due to the impact of weathering and leaching, resulting in reservoirs with pores, caves, and fractures that developed to provide a good space for oil and gas enrichment. (iii) The geological development history has seen strong tectonic movements across multiple ages. During these tectonic movements, the uplifted areas are often areas where stress is concentrated and deformation is relatively strong. Therefore, folds and faults on the uplifts are relatively developed to form various types of traps and trap combinations, providing good conditions for the enrichment of oil and gas. The Paleozoic mainly saw buried hill traps, stratigraphic unconformity traps, fault block traps, and internal anticlinal traps, while the Mesozoic and Cenozoic mainly saw drape anticlinal traps, fault traction anticlinal traps, overlying

unconformity lithologic pinch-out traps, and sandstone lens traps. It can be seen that paleo-uplifts are the most conducive to the enrichment of oil and gas, and should be the key area for future oil and gas exploration.

(2) **Paleo-slopes:** The cratonic depression basins at different stages of the Paleozoic Era were superimposed and compounded to form multiple paleo-structural slopes, such as the Maigaiti and Kongquehe Slopes. For a long time, these slopes were located in the transitional structural positions between oil-generating depressions and uplifts. Not only do they have relatively rich oil and gas resources but also they are also adjacent to oil-generating depressions. The oil and gas generated in the depressions must migrate to the slopes at a higher position. If there are good traps, oil and gas fields will form. A typical example is the first major breakthrough in the Carboniferous System was the discovery of the Bashtuo Structure of the Maigaiti Slope in 1992.

(3) **The unconformity surfaces:** Controlled by regional unconformity surfaces, most of the oil and gas is enriched in the adjacent layers above and below the regional unconformity surfaces. In particular, the first layer above the regional unconformity surfaces is an important layer for oil and gas enrichment. Examples include the Tahe, Tazhong 4, Shixi, and Donghetang Oil Fields.

(4) The high-yield oil and gas fields developed in the fault zone are closely related to the fault activities. Experience has proven that oil and gas are enriched in the layer where the fault breaks happens and that the vertical migration and enrichment height of oil and gas are completely controlled by the fault. Examples include the Karamay, Dushanzi, Qiuling, Yakra, Bashto, Kela2, and Hotan River Oil and Gas Fields.

(5) **Deep layers:** In recent years, several oil and gas reservoirs (fields) have been found in the deep layers below 6,300 m in the Tarim Basin. Liquid hydrocarbon still exists at a depth of 8,408 m in the Zengsheng-1 Well of the Tahe Oil Field, and dolomite pores are well developed. High-yield oil and gas flows are found at 5,880 m in the Yongjin area of the northern slope of the Changji Depression in the Junggar Basin, which further proves that the prospects for deep exploration are optimistic.

(6) **Volcanic rocks:** Multiple oil and gas fields have been found in the Carboniferous volcanic rocks on the northwestern edge of the Junggar Basin, the Shixi area, and eastern Junggar area; oil and gas have also been found in the Permian volcanic rocks in the Tarim Basin. The Carboniferous–Permian volcanic rocks are well developed in Northwestern China, and it is expected that there will be more discoveries of oil and gas fields.

(7) **Piedmont slopes:** The slope areas of the foreland basins are where oil and gas reservoirs are mainly found. Because they are adjacent to foreland oil-generating depressions, the oil and gas generated by the depressions must migrate toward the slopes and accumulate in reservoirs under appropriate trap conditions. Large condensate gas fields have been found in the Yaha and Dalaoba structures on the southern slope of the Kuqa Foreland Basin in the Tarim Basin. In addition, the Lukeqin Heavy Oil Field has also been found on the southern slope of the Tuha Basin.

(8) **Piedmont fault fold zones:** During the formation of a foreland basin, due to the compression of adjacent orogenic zones, surface folds are formed in rows along the detachment surfaces, accompanied by thrust faults. For example, there are five rows of surface folds developed from north to south in the Kuqa Foreland Basin, namely, the Kumuglim–Tugerming, Kasantokai–Jidik, Qiulitak, Yaken, and Daluoba (Yaha) structures. Oil and gas fields have been discovered in surface structures such as the Iqikrik, Kela2, Dina2, Dawanqi, and Kelainan structures.

(9) **Piedmont overthrust zones:** There are large overthrust nappes in front of the Tianshan and Kunlun Mountains. According to magnetotelluric data, the West Kunlun orogenic zone overthrusts toward Tarim for over 100 km. Many large oil and gas fields have been discovered in similar overthrust nappes abroad. The Karamay Oil and Gas Field was found on the northwestern edge of the Junggar Basin, another oil and gas field was found under the Silurian metamorphic rocks on the southern edge of the Jiuxi Basin, and a high-yield gas reservoir was discovered under Presinian metamorphic rocks in the Pishan area, southwest of the Tarim Basin. All these indicate that the piedmont overthrust zones provide broad prospects.

(10) **Low-order torsion tectonics:** Currently, six types of torsional structures have been discovered in the basins of Tarim, Junggar, Qaidam, and

Jiuxi, such as the broom-type, echelon, inverse-S-type, λ-type, rotational torsion, and imbricate thrust faults. As a matter of fact, multiple oil and gas fields have been discovered in the abovementioned torsion structures.

To sum up, Northwestern China, controlled by multiple tectonic systems, is rich in oil and gas resources, with broad fields of exploration and great potential. It is one of the main areas where oil and gas replacement resources are found in China.

References

[1] Li Siguang. *Introduction to Geomechanics*. Beijing: Science Press, 1973.
[2] Sun Dianqing. A basic viewpoint of geomechanics for oil exploration — The viewpoint of tectonic systems. *Geology of Petroleum and Natural Gas*, 1992 13(3): 12–21.
[3] Kang Yuzhu. *Paleozoic Marine Oil and Gas Fields in Tarim Basin*. Wuhan: China University of Geosciences Press, 1992.
[4] Kang Yuzhu and Kang Zhijiang. Significant progress of geomechanics in oil and gas exploration in Tarim Basin. *Journal of Geomechanics*, 1995 1(2): 15–19.
[5] Kang Yuzhu. *Geological Characteristics and Resource Evaluation of Oil and Gas Distribution in Northwestern China*. Urumqi: Xinjiang Science and Technology Industry Press, 1997.
[6] Kang Yuzhu. *Outlook for oil and gas prospects in Northwestern China. Contemporary Petroleum and Petrochemical Industry*, 2005 13(7): 17–20.
[7] Kang Yuzhu. Conditions and explorative directions of marine giant oil and gas fields of Paleozoic in China. *Xinjiang Petroleum Geology*, 2007 28(3): 263–266.

Petroleum Control Patterns using Structural Systems

Abstract

The theory of geomechanics as created by well-known geologist Dr. Li Siguang has been applied to effectively guide China's oil and gas exploration work, leading to successful discoveries of numerous oil and gas fields. This chapter summarizes the domestic exploration efforts of several large and medium-sized basins and puts forward an understanding of petroleum control patterns using five types of structural systems, namely, the structural system hierarchical oil control pattern, structural system superposition control pattern, structural system complex control pattern, structural system multistage control pattern, and low-level shear structures control pattern. It is believed that such an understanding is important to guide current and future oil and gas exploration in China.

Keywords: Structural system, petroleum-bearing basin, and oil–gas distribution regularity.

For more than half a century, petroleum geologists have achieved significant success in oil and gas exploration practices by applying the theory of geomechanics [1]. After a breakthrough in oil and gas exploration in the Paleozoic marine carbonate rocks in the Sichuan Basin in the 1950s, the Tahe Oil Field, the first giant Paleozoic marine oil field in China, was discovered in the Tarim Basin in the 1980s. It is also one of the largest oil fields discovered in the Ordovician System in the world, with proven reserves reaching the equivalent of $12 \times 108t$ oil in 2012. Since then, 20 oil and gas fields have been successively discovered in Xinjiang, representing the second major leap in China's oil and gas exploration. Based on the abovementioned research and practices, this chapter proposes five petroleum control patterns using structural systems [2].

1. Hierarchical Petroleum Control using Structural Systems

1.1. *Giant Structural Systems Control the Generation and Evolution of Oil- and Gas-bearing Basins*

The formation and evolution of the Songliao Basin, the Bohai Bay Basin, the Jianghan Basin, the Ordos Basin, and the Sichuan Basin in eastern China have mainly been controlled by the Neocathaysian tectonic system, the latitudinal tectonic systems, and the Cathaysian tectonic system, while the Tarim Basin, the Junggar Basin, the Tuha Basin, the Qaidam Basin, the Hexi Corridor Basin, and other basins in the north of Xizang in western China have mainly been controlled by the Western Region tectonic system, the latitudinal tectonic systems, and the Qinghai–Xizang η-type tectonic system [2, 3] (Table 1).

(1) **Source areas:** The hydrocarbon source regions of various basins in Eastern China that developed in the context of regional tensile stress

Table 1.   Major oil- and gas-bearing basins in China controlled by giant tectonic systems.

Songliao	✓		✓			
Bohai Bay	✓		✓			
Jianghan	✓		✓			
Ordos	✓	✓	✓			✓
Sichuan	✓	✓	✓			
Tarim	✓			✓	✓	
Junggar	✓			✓		✓
Qaidam	✓			✓	✓	
East China Sea	✓		✓			

are all controlled by primary subsidence zones, such as the Zhedong Depression and Wendong Depression in the East China Sea Basin, the central depression in the Songliao Basin, the Jiyang–Huanghua Depression and Beizhong Depression in the Bohai Bay Basin, and the Qianjiang Depression in the Jianghan Basin. The hydrocarbon source regions of various basins in Western China that developed in the context of regional tensile stress, however, are mostly superimposed by multiple negative structural systems, resulting in multiple layers, high thickness, and wide planar distribution of hydrocarbon source rocks on the profile. Examples include the western and southern depressions of the Ordos Basin, the depression areas of the Sichuan Basin and the western foreland basin, the central depression of the Junggar Basin, the northern Tianshan Foreland Basin on the southern edge, and the southwest depression and northeast depression of the Tarim Basin.

(2) **Primary uplift and slope zones (areas) control oil and gas enrichment zones:** As mentioned, the primary subsidence zones in the basins control the hydrocarbon source areas, so the primary uplift and slope zones (areas) adjacent to the hydrocarbon source, as well as the uplift and fault zones within the oil-generating depression, first become oil and gas accumulation areas. Typical examples include the Daqing Placanticline, the Liaoxi Slope, the Central Uplift and Slope Zone of the Bohai Bay Basin, the Yishan Tectonic Zone, the Central Paleo-uplift of the Ordos Basin, the Paleo-uplift and Slope Area formed during the Indosinian and Yanshan Movements in the Sichuan Basin, the Central Uplift, Santai Uplift, and Northwest Edge Thrust Zone of the Junggar Basin, the Tazhong Uplift, Shaya Uplift, and Maigaiti Slope of the Tarim Basin [4], and the Laojunmiao uplift zone in the Jiuxi Basin.

1.2. *Primary Structural Systems Control Hydrocarbon Source Areas and Hydrocarbon Accumulation Zones*

(1) **Primary subsidence zones control the distribution of hydrocarbon source areas:** The hydrocarbon source regions of various basins in Eastern China developed in the context of regional tensile stress are all controlled by primary subsidence zones, such as the Zhedong Depression and Wendong Depression in the East China Sea Basin, the central depression of the Songliao Basin, Jiyang-Huanghua Depression

and Beizhong Depression in the Bohai Bay Basin, and Qianjiang Depression in Jianghan Basin. The hydrocarbon source regions of various basins in Western China developed in the context of regional tensile stress, however, are mostly superimposed by multiple negative structural systems, resulting in multiple layers, high thickness, and wide planar distribution of hydrocarbon source rocks on the profile. Examples include the western and southern depressions of the Ordos Basin, the depression areas of the Sichuan Basin and the western fore-land basin, the central depression of the Junggar Basin and the north-ern Tianshan foreland basin on the southern edge, and the southwest depression and northeast depression of Tarim Basin.

(2) **Primary uplift and slope zones (areas) control oil and gas enrich-ment zones:** As mentioned above, the primary subsidence zones in the basins control the hydrocarbon source areas, so the primary uplift and slope zones (areas) adjacent to the hydrocarbon source, as well as the uplift and fault zones within the oil generating depression, first become oil and gas accumulation areas. Typical examples include the Daqing Placanticline, Liaoxi Slope, Central Uplift and Slope Zone of Bohai Bay Basin, Yishan Tectonic Zone, Central Paleouplift of Ordos Basin, Paleouplift and Slope Area formed during Indosinian and Yanshan movements in the Sichuan Basin, Central Uplift and Santai Uplift and Northwest Edge Thrust Zone of the Junggar Basin, Tazhong Uplift and Shaya Uplift and Maigaiti Slope of Tarim Basin [4], and Laojunmiao uplift zone in the Jiuxi Basin.

1.3. *Secondary Torsion Structural Systems Control the Distribution of Oil and Gas Fields*

As a result of the different generation and development processes of the secondary structural systems in the basins, their control over oil and gas varies greatly, and can generally be divided into three types.

(1) The various levels of direct torsion structures within the torsion struc-tural systems are often distributed in a zigzag pattern. The negative zigzag-type structures control the distribution of oil-generating depres-sions, while the associated positive structures are the main places for oil and gas accumulation. However, this pattern of oil and gas distribu-tion is often transformed by later secondary structural systems.

(2) In the middle and late development stages of the giant tectonic systems, due to changes in boundary conditions and stresses, the rotational and torsional stresses are often superimposed on the background of the direct torsional stress field. When the vortices overlap in the early vertically twisted negative depressions, the subsidence amplitude and range of the depressions are further expanded and become good oil-generating depressions. As a result, the arcuate cyclic structural zones become rich zones of oil and gas.

(3) Negative structures formed by small-scale torsional structures, whether through either twist or torsion, which are small in scale and generally do not have oil-generating condition. Only when they are developed in or near oil-generating depressions can they have the opportunity to accumulate oil and gas. When they are far from the oil-generating depressions, it is difficult to enrich oil and gas, even if the torsional structures are well developed.

2. Superimposed Petroleum Control using Structural Systems

The tectonic systems evolve in distinct stages that are characterized by different activities. For example, the Manjiaer Depression in the northeastern part of the Tarim Basin was formed by three phases of subsidence, i.e., during the Sinian–Ordovician, the Silurian–Devonian, and the Carboniferous–Triassic, so the depression became a multi-phase oil source area. From the Late Ordovician, the Tazhong Uplift became an NE-trending uplift of the Western Region tectonic system and underwent a Silurian–Devonian inherited uplift to form a superimposed uplift, thus becoming a major oil and gas enrichment zone, where multiple oil and gas fields have been discovered [4].

In addition, the Junggar Basin, the Ordos Basin, the Sichuan Basin, and other basins all have the same tectonic system which is formed by multiple periods of superimposed petroleum control.

3. Composite Petroleum Control using Structural Systems

Existing data indicate that all major basins were generated and evolved under the combined control of multiple tectonic complexes. For example, the piedmont depression on the western edge of the Ordos Basin was mainly caused by the composite action of the Cathaysian tectonic system and the Helan Mountain ε-type tectonic system. Therefore, three sets of hydrocarbon source rocks developed in this depression (Cambrian–Silurian, Carboniferous–Permian, and Triassic–Jurassic), which became an important oil and gas source area. The central part of the basin has been in a relatively uplifted state from the Cambrian to Mesozoic and has become an oil- and gas-enriched area (zone) [5].

In the Sichuan Basin, under the joint control of the Cathaysian and the latitudinal tectonic systems during the Paleozoic, the eastern and southern parts of the basin deposited during the Cambrian–Silurian were important oil source areas. The central part of the basin is located in a relatively uplifted area, and the uplifts in different periods have become favorable locations for oil and gas enrichment, such as the Weiyuan Gas Field and the Puguang Gas Field.

In addition, the Naxu gas field in southern Sichuan is a typical example of a gas field controlled by different structural systems (Figure 1). The Naxu gas field is a composite structure formed by the action of longitudinal and latitudinal horizontal pressures in the Naxu chessboard zone. In the early stage, the main structure distributed in the east–west direction was formed by the action of meridional pressure. In the late stage, it experienced a relatively horizontal pressure in the latitudinal direction,

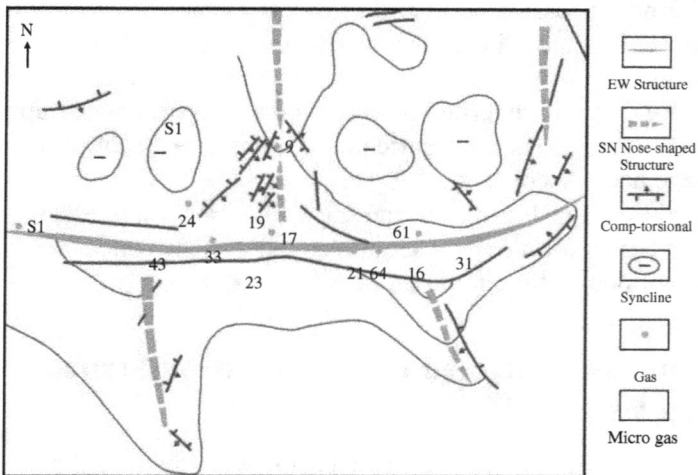

Figure 1.   Structural features of the upper boundary of the lower Permian in the Naxu Gas field, Sichuan Basin.

resulting in a north–south nose-shaped structure on both wings of its east–west part. The intersection of the nose-shaped structure and the east–west structure is precisely the culmination of the combination between the east–west and north–south structural systems. The location the corresponding to the north–south nose of Yanggaosi is the primary part of the culmination, while the two secondary parts of the culmination in the east and west correspond to the south and north nose structures, respectively [5].

Due to horizontal pressure, this type of gas field features poor development of tensile fractures at the top of the fold, and inconsistency in the distribution, density, extension length, and fracture width of the remaining tensile fractures. During drilling, oil and gas communication channels are limited, and local scale connectivity is often formed along the direction of unevenly developed longitudinal or transverse fractures. Gas wells are mainly distributed in and around complex locations.

4. Characteristics of Structural Systems with Multi-stage Hydrocarbon Accumulation Resulting from Multi-stage Activities

A variety of structural systems feature multiple stages of activities, and the generation and final formation of each structural system are not caused by a single tectonic movement. Each activity of each relevant structural system has a direct impact on oil and gas generation, hydrocarbon expulsion, and reservoir formation.

For example, there were three periods of hydrocarbon expulsion and accumulation in the Tahe Oil Field in the Tarim Basin: the Hercynian Movement Period, the Indosinian–Yanshan Movement Period, and the Himalayan Movement Period. There were also three stages of reservoir formation in the Sichuan Basin: the late Hercynian Movement, the late Indosinian Movement–early Yanshan Movement, and the late Yanshan Movement–Himalayan Movement [6–9].

5. Control of Torsional Structural Systems Over Oil and Gas Distribution

The author has proposed six hydrocarbon control models for torsional structural systems, i.e., broom-type, echelon-type, rotational torsion, λ-type, S-type, and imbricate structural systems [5].

5.1. *The Enrichment Degree of Hydrocarbon Fluids in Rotational and Torsional Structures Features Certain Selectivity*

Analysis of the oil- and gas-bearing characteristics and stresses of various rotational torsional structures reveals that rotational torsional stresses are the driving force for the hydrocarbon fluids to flow in rotational torsional structures, from high-stress areas to low-stress areas. Therefore, identifying the source and transmission direction of rotational and torsional stresses is necessary for determining the origin and nature of rotational and torsional structures (active or passive), as well as the distribution of oil and gas in them.

Generally speaking, the degree of oil and gas enrichment gradually decreases from the inner cycle layer to the outer one. The degree and location of oil and gas enrichment depend on the mode of stress action and the direction of oil and gas migration and accumulation. If the center of a rotational torsional structural system is a spiral structure of an oil-generating depression, the degree of oil and gas enrichment increases from the convergence direction to the spreading direction. If it is a spiral structure of a pillar, the degree of oil and gas enrichment increases from the spreading direction to the convergence direction. For broom-type structures, the migration and accumulation direction and enrichment location of oil and gas should be generally consistent with those of spiral structures. For inverse-S-type structures, due to the stress coming from the periphery, oil and gas are mostly enriched in the north and south ends and the middle zone, especially in the areas with the largest bending rate.

5.2. *Fractured Structures with Different Properties have Different Control Effects on Oil and Gas*

Analysis of the fracture surfaces finds that the compression torsional fracture surfaces feature good sealing performance. Therefore, under favorable hydrocarbon source conditions, primary oil and gas enrichment is likely to occur on the near source side, and the closer the fault is, the higher the oil and gas productivity will be. Compression-torsional fractures are also natural barriers for oil and water. During development and water injection, oil fields are often divided into different units. The

Xijiakou Oil Field in the Jianghan Basin; the Daqing Oil Field, the Xingxi, and the Aogula Reservoirs in the Songliao Basin; and the Tuozhuang–Shengli Village Oil Field in North China are all good examples (according to Zhang Fuli).

Tension-torsional fracture structural planes with good opening properties are often used as oil and gas transmission channels. Oil and gas reservoirs can be formed when there are traps on the hanging wall of a fault or in the stratum through which fault intersects.

It should be noted in particular that the relationship between fracture structural planes and oil and gas is also controlled by the lithological conditions on both sides of the fracture structural planes, igneous rock activities, and other conditions.

6. Geostress is the Main Driving Force for Oil and Gas Migration and Accumulation

Geostress not only acts on the strata to form various structural features but also acts on the fluids in the rock mass, which are much more sensitive to the same geostress than rocks. Therefore, the migration laws of oil, gas, and water can also reflect the characteristics of geostress activities in the region.

For example, the actual inclination direction of the oil–water interface in the Daqing Oil Field is opposite to the inclination direction caused by the hydrodynamic drive, but consistent with the inclination direction caused by stress. The oil–water transition zone under the pure oil-bearing zone in the Putaohua Oil Field is thick at the structural axis and thin near its wings, showing an upward curved crescent shape, which indicates that the oil–water gravity differentiation is still ongoing. Only when the geostress that causes oil–gas migration and accumulation and forms structural patterns is basically eliminated can gravity function to achieve a new equilibrium in the oil, gas, and water differentiation in the trap.

It can be seen that oil and gas migration and accumulation are controlled by a certain stage of evolution of the relevant structural system. That is, the period of intense geostress activities when the structural system forms is the main period of oil and gas migration and accumulation. Due to the volatility of such geostress activities, the oil and gas migration and accumulation are episodic.

References

[1] Li Siguang. *Introduction to Geomechanics*. Beijing: Science Press, 1973.

[2] Kang Yuzhu. *Main Tectonic Systems and Oil and Gas Distribution in China*. Urumqi: Xinjiang Science and Technology Press, 1999.

[3] Kang Yuzhu. Tectonic systems in Northwestern China and their relations with hydrocarbon. *Xinjiang Petroleum Geology*, 2009 30(4): 407–411.

[4] Kang Yuzhu and Wang Zongxiu. *Research on Petroleum Control by Tectonic Systems in the Tarim Basin*. Beijing: Geology Press, 2009.

[5] Kang Yuzhu and Wang Zongxiu. *Research on Petroleum Control by Tectonic Systems in Northwestern China*. Beijing: Geology Press, 2012.

[6] Kang Yuzhu, Zhang Ximing, Kang Zhihong, *et al. Tahe Oil Field in Tarim Basin, China*. Urumqi: Xinjiang Science Press, 2004.

[7] Zhou Xinyuan, Li Benliang, Chen Zhuxin, *et al.* Tectonic genesis and exploration direction of Tazhong oil and gas field. *Xinjiang Petroleum Geology*, 2011 32(3): 211–217.

[8] Wang Weiguang, Lv Xiuxiang, Yu Lian, *et al.* Tectonic pivot zone of the Tarim Basin and its hydrocarbon accumulation conditions. *Xinjiang Petroleum Geology*, 2011 32(4): 333–337.

[9] Wu Guanghui, Ju Yan, Yang Cang, *et al.* Control of tectonic systems over Ordovician Reef Beach type reservoirs in Central Tarim basin. *Xinjiang Petroleum Geology*, 2010 31(5): 467–470.

Index

www.ingramcontent.com/pod-product-compliance
Lightning Source LLC
Chambersburg PA
CBHW050559190326
41458CB00007B/2102